Thinking of...

Building a smart utility?

Ask the Smart Questions

By Dirk Michiels

Smart Questions™ Philosophy

Smart Questions is built on 4 key pillars, which set it apart from other publishers:

1. *Smart people want Smart Questions not Dumb Answers*
2. *Domain experts are often excluded from authorship, so we are making writing a book simple and painless*
3. *The community has a great deal to contribute to enhance the content*
4. *We donate a percentage of revenue to a charity voted for by the authors and community. It is great marketing, but it is also the right thing to do*

www.smart-questions.com

Reviews

Adacta

Adacta is with its 250 employees in Slovenia, Croatia and Serbia the leading Microsoft Dynamics partner in the Adriatic region. We often act also as a global implementation partner delivering international projects in Central and Eastern Europe.

Energy markets in our region were in the last few years undergoing significant deregulation changes imposing new challenges and opportunities to traditional market players and to new entrants. New business paradigm requires a modern, flexible and cost-efficient IT support with functionalities exceeding the level that can be reached simply by reasonable additional investments in the legacy applications.

Microsoft Dynamics AX provides a great platform to develop industry-specific vertical solutions and we are amazed by modern concepts and past experience built in the MECOMS™ solution. Flexibility of the AX and MECOMS™ architecture allowed us to support efficiently also conceptual requirements that are specific for our markets.

This book offers a rarely seen combined overview of business and IT concepts in the utility sector and I would recommend it to everybody involved in IT-related decisions and projects in the utility industry.

Aleš Zajc, Sales and Business Development Manager, Adacta programska oprema d.o.o., Slovenia

Avanade

Avanade is the Joint Venture of Microsoft and Accenture, specialized in implementing and optimizing the Microsoft solutions for the Enterprise.

With 17.000 professionals, Avanade operates globally, and are among the 3 largest Dynamics AX system integrators in the world, helping our clients to deliver results in the design, implementation roll-out, operation and application management of agile ERP systems, using 'best in class' Business Process models inherited from years of AX implementations and Accenture experience.

Focusing on innovation and technology to better address business challenges in the moving utilities world, we have dedicated specific teams and investments in the Utility space for years, and have developed an end to end commodity trading capability that many global or local utilities are leveraging

The Avanade/Accenture/Ferranti Alliance and value proposition

MECOMS™ offers unprecedented system flexibility to allow utilities to act smart, and anticipate on the major business transformation challenges ahead. Avanade selected MECOMS™ to be our unique industry solution for smart utilities, building on top of our very strong Dynamics AX capabilities and integrating the full Microsoft technology stack (portals, BI, EAI...) into a unified solution.

AVANADE wanted an alternative to the more classic heavy weight approach to CC&B for utilities, to address quick moving companies, in the most appropriate models: either full ERP for Utilities, or a more light weight solution in a 2Tier approach complementing an existing SAP or Oracle ERP, with handling only specific business workloads within AX/MECOMS.

After initial successful collaboration with Accenture and FERRANTI in implementing MECOMS, AVANADE and FERRANTI shaped a Global Alliance in 2011, by which the companies collaborate in all geographies to bring to market the most advanced and agile customer care, billing and meter data management solution: MECOMS™.

This Alliance also includes our mother company Accenture who needs to be able to deliver MECOMS™ solutions for its clients, leveraging AVANADE as its AX arm.

More than a system feature only... "agile" is also defining the way the market is expecting us to behave, as solution providers and implementers. Avanade has the right size, the obsession on client satisfaction, utilities industry experience and assets installed at the largest customers, and global methods that allow us to deliver results consistently.

François Matte, Avanade Global

CGI

CGI is a leading global information technology and business process services company that offers consulting, systems integration, outsourcing and innovative solutions to gain operating efficiencies, reduce costs and improve competitive advantage. CGI has always been a reliable and trusted partner committed to our clients' success, providing flexibility, accountability and quality through our client proximity business model and the application of our Management Foundation.

Through deep industry and domain expertise and a comprehensive portfolio of consultancy, business solutions and outsourcing services, we enable our Utilities clients to succeed in both regulated and competitive markets around the world. We have more than 35 years of experience in the sector along with a proven track record that position us as one of the leaders in it services for Utilities.

Our value propositions cover our clients' supply chains from production/generation through distribution to customer relationships. Our focus is directed by the major trends in the sector and we provide a full portfolio of solutions and services including the full management of IT and business functions, business and IT consulting, systems integration, and a field-proven portfolio of software solutions.

CGI (formerly Logica) has been a MECOMS™ partner for the last 4 year and in that period MECOMS™ has become a strong asset in our global Utility portfolio. In CGI we have managed to build new very strong value propositions for Utilities both in shape of business solution implementations as well as a new innovative Business Process Services concept with MECOMS™ as the foundation.

With our BPS concept for Utilities we can offer full or partial outsourcing of all Front and Back Office processes such as billing service, service desk and automatic meter reading in a flexible and scalable manner based on MECOMS™ as the future-proof and user-friendly industry platform. The Utilities BPS concept is made possible through MECOMS™ combination of flexibility, efficiency and full range of support for all Utility business processes.

In a world of constant change and with ever increasing pressure on the Utility sector, MECOMS™ is in my opinion the best and only integrated ERP, CIS and MDM platform for energy- & utility companies. MECOMS™ is a truly innovative, flexible and technically state of the art solution that allows customers to implement MECOMS™ as a focused solution for specific business areas such as MDM, CRM and market interaction or as and complete end-to-end solution for all processes.

Mads Brøgger, Markedschef Utility, CGI Danmark

Euriware

As a leading Information Systems provider for Energy, Industry & Defense markets in France, providing our customers with large capabilities to conceive, construct, maintain and exploit highly critical systems, we do need to choose partners who share the same values as ours. Ferranti is one of our first class Software partner giving us the opportunity to build together the smart business enablers our customers need in the Energy and Utilities market.

One more time, Ferranti give a vision we share with our customers in the utility market.

In this book, Dirk provides us a deep perspective of all the fields in the energy and utility market, of its different issues, ask the right - and smart- questions, and a complete view of the multiple viewpoint of implementations in this sector.

It also gives us a complete review of the MECOMS™ solution associated with Microsoft technologies, meanwhile completed with the right approach, reminding us of the importance of "the fabulous six".

As a consequence, we, at Euriware, are convinced this book can be read and used at multiple levels for readers involved at different levels of knowledge in the energy and utility sector. A must have. A smart book.

Patrick Clovirola, Directeur Offres et Alliances, Division Intégration de Systemes, Euriware France

Giza Systems

Giza Systems *www.gizasystems.com* is the number one systems integrator in Egypt and the Middle East providing a wide range of industry specific technology solutions in the Telecom, Utilities, Oil & Gas, and Manufacturing industries. Since 1974, our consultancy practice has provided industry focused services that enhance value for our clients by streamlining operational and business processes. Operating in the Middle East through Giza Arabia www.gizaarabia.com, Giza Systems Qatar and Giza Systems JLT (UAE), our group of companies is focused on contributing to the local and regional development with our technology solutions, commitment and outstanding customer service.

Michiels' description and explanation of the MECOMS™ solution clearly and concisely answers the question of implementing smart solutions for smart utilities. The MECOMS™ solution offers a realistic and workable method for providing successful customer care, billing, and meter data management. The models presented in Michiels' book are extremely informative and his detail regarding project approach and data migration is an asset to anyone considering the MECOMS™ solution.

Osama Sorour, Vice President Business Development, Giza Systems Middle East and Africa

GMCS

"Economic and social challenges in global world require changing of how managers of utility companies are thinking, and communicate each other with help of IT technologies. This book is a must-read for those who is looking for pragmatic approach overcoming challenges utility companies are facing now. Matters, discussed in this book, will guide through a maze of words about smart utility before reader will find individual practical way of how IT can do utility company smarter."

Oleg Lysov, Vice president GMCS, Russia

NCS

NCS is a leading info-communications technology (ICT) service provider in Asia-Pacific region and has presence in 22 countries located throughout APAC, Europe and the USA. NCS delivers end-to-end ICT and communications engineering solutions to help governments and enterprises realise business value through the innovative use of technology. Our unique delivery capabilities range across consulting, development, systems integration, outsourcing, infrastructure management & solutions and digital engagement. We also provide mobility, social networking, business analytics and cloud computing services.

This book provides a candid, insightful and comprehensive look at how to implement more intelligent solutions for the energy and utilities sector. MECOMS™ solution strongly aligns with NCS' focus on bringing great value to customers. It enables utilities of any size a head start on how to streamline information, businesses and processes with state-of-the-art technology effectively and efficiently.

Chia Wee Boon, CEO NCS Pte. Ltd. (A member of the SingTel Group), Singapore.

Praxis

We at Praxis, creating the COE for the Utility Industry for extending our core strength in Dynamics AX and MECOMS™ solution offering. We believe the Energy & Utility Industry is rapidly growing and changing Industry and having an enormous opportunities for IT /ITES companies. Being a System Integrator, we are keen on increasing our strength in delivering the right solutions to Energy & Utilities Segment. We are successful in delivering our first implementation with MECOMS™ at Electric Utility Organization in India.

MECOMS™ is the ever Best Solution for Utility Industry, which is rapidly penetrating into industry because of Solution Rich functionality and Flexible operation agility. In my view MECOMS™ is the only solution which gives complete 360° information of any Utility Organization.

Hari Krishna, CEO Praxis, India

Prodware

Founded in 1989, PRODWARE creates, integrates and hosts IT solutions for businesses.

Prodware acts in a sector where success requires a combination of strong IT expertise and industry know-how.

PRODWARE serves more than 17.500 active clients, and is the key actor and partner for the installation and management of global IT solutions and applications.

MECOMS™, smart solution for smart utilities packages best practices and utilities knowhow together with a great software.

As the utilities and energy business is growing rapidly, customers need an agile solution to enable them to grow and adjust to the frequently changing energy market.

For people from the utilities and energies industry, this book is inspiring.

Yossi Haimov, Managing Director, Prodware Israel

Sterlite

Sterlite Technologies Limited is a leading global provider of transmission solutions for the telecom and power industries.

Equipped with a product portfolio that includes optical fibers, telecommunication cables, solutions for system integration, fiber connectivity solutions, neutral networks, power cables and a comprehensive power conductors portfolio, Sterlite's vision is to "Connect every home on the planet".

Sterlite is also executing multi-million dollar power transmission system projects, pan-India.

"Thinking of...Building a Smart Utility" is a must read for all Utility IT professionals. In today's utility- world, professionals are under extreme pressure to do more in less time and with fewer resources. Against this background, this is a smart initiative to cover succinctly but insightfully a range of themes spanning from market mechanics to informational needs to project implementation ethos to passion of utility people. Internationally recognized as a thought leader, Dirk Michiels, in his inimitable style has depicted to all of us what should be done to graduate as a successful utility-IT professional."

Sukhminder Singh Sidana, Head– Business Development, Sterlite Technologies Limited, India

WBI

WBI is on Slovakian and Czech market one of the most important partners for Dynamics solutions of Microsoft. With more than 140 customers, localization partnership with Microsoft for Microsoft Dynamics NAV 2013 and AX 2012 R2 for Slovakia and rapid growth in year 2012 WBI reached the membership in prestigious Inner Circle Club of Microsoft Dynamics worldwide partners.

WBI's success and strategy is based on 4 pillars: Innovations and complexity, Experienced team of professionals, Long-term satisfaction of our customers and Company culture

MECOMS™ solution and Partnership:

WBI is relatively young partner of Ferranti and its solution MECOMS™ for utility. The choice of this vertical and the partnership with Ferranti is based on the dynamic changes of the market with utilities and the de-regulation of the market. These changes call for Smart or Smarter (like Dirk Michiels mentioned) Solutions. I appreciate this great book which covers both – information about actual situation on the market with the utilities and the right approach to building a solution based on MECOMS™ as well as a large amount of aspects and open questions relevant for all types of companies operating on the utility market.

Dita Sirotová, Sales Director WBI, s.r.o., Slovakia

Author

Dirk Michiels

Dirk Michiels is the CEO of Ferranti Computer Systems. He is responsible for the strategic and operational leadership of Ferranti worldwide, covering sales, marketing, services and product/solution development. Michiels leads the worldwide organization offering the MECOMS™ solution for the Energy and Utility market and the regional ICT Infrastructure and System Integration business.

Ferranti Computer Systems N.V. is a company of the Nijkerk group.

Dirk Michiels grew up in Belgium near the city of Antwerp and holds a master degree in Electronics and Chip Design. He joined Ferranti in 1986 at the Ferranti Wythenshawe division in Manchester UK. He worked in development, system architecture and project management of the civil systems group focusing on the energy, oil and gas and process industry. He worked on various projects for Shell.

With the deregulation in Europe in the mid 90's, Michiels got very much involved in defining processes and business models for utilities companies. This has inspired and fuelled the development of the MECOMS™ product.

As of 2005, Dirk Michiels has been driving, as Chief Operations Officer, the MECOMS™ product development on the Microsoft Dynamics platform and the relationship with Microsoft, partners and analysts.

He is also president of IAMCP (International Association of Microsoft Channel Partners) for the BELUX Chapter.

Dirk now lives in Belgium with his wife. Besides work, he spends time on reading, enjoying food, cooking and hiking in Scotland.

Table of Contents

Acknowledgements

My passion is solving business challenges using "people energy".

I strongly believe people are the determining factor to offer innovative solutions to the challenges companies are faced with.

This is reflected not only in products and solutions but also in the way people work together to build and implement systems. This is what gives me job satisfaction and makes work enjoyable.

The Energy and Utilities market is changing significantly. This impacts IT systems and the way we build and maintain them in a way that is so drastic we like to call it "the storm is coming".

With this book you get access to thousands of years of experience within Ferranti and its partner community in building solutions for the Energy and Utility space using the MECOMS™ solution on Microsoft Dynamics AX.

I could not have written it on my own. Many people have contributed to give you a condensed 360 degree view on aspects related to implementing systems in the Energy and Utilities market. I would like to thank:

- The MECOMS™ Leadership team, Guido Van de Velde, Tom Van Haute, Johan Vandekerckhove, Geert Janssen, Johan Crols, Marc Crauwels for the support and insights in structuring and reviewing this book. It is their thought leadership and years of experience that makes MECOMS™ a world class product. Thibaut Vos, Barbara Peeters and Kaori Konta for providing valuable input. Also, a word of thanks to Martina Vroblova for her passion and continuous effort in improving our implementation practices. Alain Bastiaens for his people feel. Vincent De Munck for graphics. Pieter Mestdagh and Anne Willemse for editing. Wim Mangodt and Biswapriya Mukherjee for their much appreciated input. A very special thanks to Franky Baert, my mentor, for his long term view. There would not have been a MECOMS™ product without his vision.
- The Ferranti solution advisory groups and project teams – worldwide for providing input.

- The Microsoft Business Solution teams for the real team work to make the product a success. To the Microsoft Enterprise Utilities group for their help in positioning the solution on a worldwide scale.
- The partners, who give us the reach worldwide to do quality implementations for utility customers.
- The Ferranti Computer Systems board of directors and its chairman Richard Nijkerk – for their strategic guidance and energy.
- Thanks to all whom I have forgotten, colleagues, consultants, analysts, all the customers, prospects and friends for their insights and support.
- Last but not least thanks to my wife, Ria, for all her patience and support in writing this book.

Dirk Michiels, the author.

Foreword

Jon Arnold, Worldwide Director Power and Utilities at Microsoft

It is with great pleasure that I can recommend this effort by Ferranti Chief Executive Office Dirk Michiels and his incredible team at Ferranti Computer Systems.

As the Worldwide Director of Microsoft's Global Power and Utilities Practice, I've had the privilege of working with Michiels and Ferranti Computer Systems because they are partners to Microsoft, and its MECOMS™ 2012 Solution, offering Customer Information System and Meter Data Management, has been recognized as "Certified for Microsoft Dynamics." Such designation doesn't come easily and it speaks to their commitment, with Microsoft, to the global utilities industries.

Now, with this book, Michiels endeavors to dig deeper into the mechanics of transporting commodities to customers and then converting the resulting meter reading into an invoice and collection process. It sounds so simple. And yet, with everything involved for millions of customers it's incredibly complex, especially given the heightened expectations of consumers and the unprecedented pace of change in the worldwide utilities business. This book, like Microsoft's Smart Energy Reference Architecture, accepts the basic premise that so much is new it needs a fundamental rethinking as to the basic mechanics for delivery and execution.

For its part, Microsoft contributes its Dynamics AX solution to Ferranti MECOMS™. At its most basic function, Dynamics enables our partners to help their utility clients manage the huge volumes of customer and operational data that are generated each minute in today's information technology driven operating environment. Dynamics works by managing data for customer relationship management and enterprise resource planning activities, and dispersing that data throughout the utility in a meaningful way. Ferranti's MECOMS™ solution ties it all together in the utility context through its Energy and Utilities Community Architecture; a modular data model built to hand all components of performance management, like customer management, meter data management, interaction management and enterprise asset management.

This book is must reading because it provides insight into the robust design and architecture that we are engaged in with Ferranti. As a blueprint for smart grid and smart energy delivery and billing operations utilities can view the competitive advantages they gain from this partnership and the resulting MECOMS™ solution. Not only does our partnership provide an integrated technology stack allowing cost effective implementation of future and responsive changes but the architecture enables Ferranti to incorporate in MECOMS™ the benefits of the billions that Microsoft invests every year in Research and development. Ultimately, our partnership delivers to utilities and their customers the new and continuously innovating benefits of cloud, mobile/devices, social and information. Other commercial Customer Information Systems will always struggle to re-architect their solutions to address the needs of utilities given the market and technology trends, and the advent of the new energy consumer. By contrast, Ferranti and Microsoft have already invested in a solution that is designed to quickly leverage new capabilities available from cloud, mobile/devices, social and information/Big data computing services. Survivors will need the technology agility that most quickly and appropriately adapt to customer, industry and technology changes.

So let's get to the core of the reason this book will be useful. Utilities companies find themselves in dire need of better data management processes because of the complex challenges they face each day. For instance, they are increasingly subject to mergers, acquisitions and privatizations, with many former monopolies having been forced to pursue new models of value creation. However, having gained the capacity to develop in other sectors and geographies, many utilities have also expanded their spheres of operations. With ever increasing retail competition, more demanding customers and further de-regulation (liberalization), energy and utilities companies must respond with speed, foresight and informed approaches to keep their costs low, and their operations humming. The graphic below summarizes the market, technology and business technology challenges that solutions like MECOMS™ and Dynamics AX must address.

Market Trends	Technology Trends	Business & Technology Challenges
Increased focus on customer focus, impeccable service	Communication technology	New business models; added value products & services
Financial crisis fallout	Mobile technology	Business process efficiency
Renewable power sources	Business intelligence	Speed to market with operating agility
Environmental concerns	Analytics	Achieving unattended behavior focused solely on exceptions
Security	Web 2.0	Workflow driven processes for "First Time Right" low costs
Mergers, acquisitions and divestments	Electrical vehicles	Clean, service oriented Architecture for simple, modular IT landscape
Aging assets	Energy storage	Flexible design for initial implementation, future change
Aging workforce	Advanced metering infrastructure	Integrated technologies for cost effective change implementations
Smart grid	SaaS	
Energy technology consumption	Cloud computing	
	IT and OT convergence	
	Demand-response.	

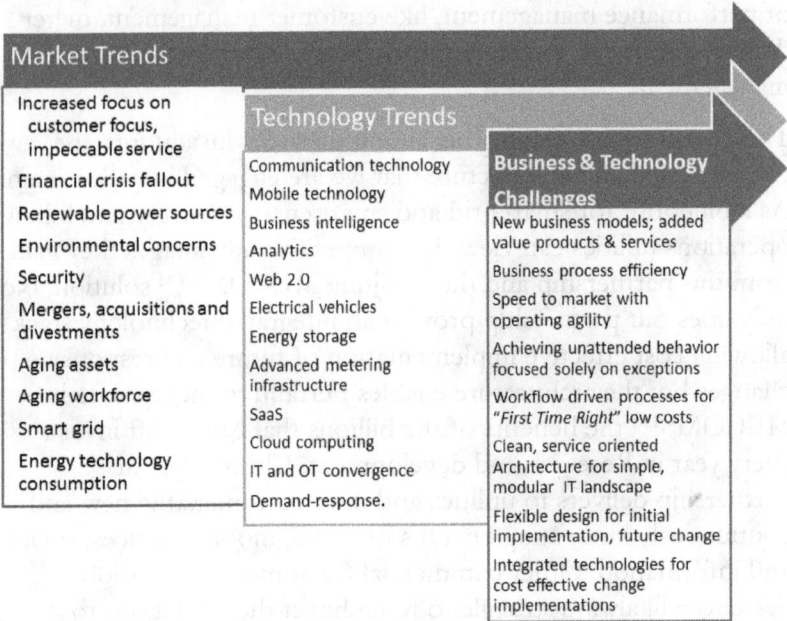

  The value of this book, "Thinking of… Building a smart utility?" is that it offers a detailed, informed, and organized vision as to how utilities can leverage the huge investments that Microsoft has made in its Dynamics™ AX, SQL Server, SharePoint communication and collaboration technologies, in partnership with the investments Ferranti has made in its MECOMS™ platform, providing problem-solving technologies for these current market and technology trends. Microsoft Dynamics AX provides utilities' employees with business tools they are familiar with, increasing their use throughout the organization, lowering business transition costs, and allowing consistent decision making throughout the organization. Dynamics' customer-oriented data model as well as its 360° cockpits enables utility employees to efficiently manage their day-to-day work and create customer satisfaction.

Greg Guthridge, Managing Director, Accenture Energy Consumer Services.[1]

Energy consumers around the world have growing expectations for energy-related products and services that align with their values and lifestyle. Smart and in-home technologies, dynamic rate structures, and emerging options from telecommunications and consumer retailers are changing the energy experience. Energy consumers now expect service, anytime, anywhere, and are seeking personalized products and services, new value propositions and increasing levels of convenience - all of which go beyond the traditional energy experience.

At the same time energy providers face rapidly evolving marketplaces and an environment of continuous change. Many energy providers are facing continuing pressures to maintain or reduce costs while also enhancing the customer experience – doing more with less. Furthermore, the traditional energy marketplace has begun to converge with other markets. This convergence is attracting new challengers vying for a share of consumer energy revenues. Even in regulated energy markets, consumers and businesses have new options as non-energy providers begin to offer products and services designed to complement the commodity, more effectively manage energy usage or even allow consumers to generate their own energy.

Many energy providers are asking how new capabilities can be introduced to capture and deliver value in response to these changes, while balancing the need to maintain or lower overall operating costs and having the flexibility to respond to a rapidly changing energy landscape. In the face of the opportunities and challenges ahead, providers will need a proven, flexible, highly-integrated IT infrastructure that leverages industry-specific leading practices to optimize customer care and billing, meter data management, operations and market interaction.

Accenture, Avanade, Ferranti and Microsoft have a strategic relationship to help our electricity clients evolve into more agile, efficient, differentiated and innovative enterprises by leveraging our business, technology and implementation experience. We work closely to help leverage the potential of cloud computing, analytics, mobility, and other emerging technologies quickly, reliably, and cost-effectively. Together we are working to make IT a strategic enabler for our clients in order to improve the overall customer experience, deliver deep customer insights, and build the flexibility needed to cost effectively meet the needs of the emerging energy marketplace.

Nigel Spooner, Vice President, Global Utilities, CGI

It is now nearly 30 years since I started in the Information Technology industry in the UK, working with a range of public sector utility companies. It seemed to us then that the sector hadn't changed much for a hundred years and didn't look as if it would change much in the next hundred. But how wrong we were! Privatisation and competition came to the UK electricity industry in the late 1980s and the changes have not stopped since.

In the ensuing years not just the UK but much of the rest of the world has seen huge developments in utilities markets, driven not only by political and commercial pressures but by responses to climate change as well. These developments have necessitated fundamental re-structuring of the utilities' operational and business models as well as very different approaches to relationships between suppliers and consumers. These relationships have also been affected over the years by new concepts such as 24/7 wireless communications and the increasing availability of real-time data, as well as the empowerment now given to consumers also to produce their own energy.

It is in the light of this wide range of complexities that Dirk Michiels' new book is both timely and hugely useful. In covering the broad spectrum of processes that are now required to ensure an effective and profitable relationship between a utility and its customers, the book adds real insight for the wide range of professionals that are involved in different aspects of that relationship. Inevitably it also highlights the essential role which Information Technology must play in enabling those processes, focusing on the need to ensure a tight combination of stable but flexible platform on which to sit rich, industry-specific function developed and delivered by real subject matter experts.

With Dynamics AX providing that platform it is impressive but unsurprising that Microsoft has chosen to promote MECOMS™ as its preferred billing solution for Utilities. It was equally natural for Logica – now part of CGI – to have chosen to become an implementer of MECOMS™ by using its broad industry knowledge to ensure that clients receive the best and most comprehensive solution to their business challenges. We believe that our many years of experience around the enablement of competitive energy markets in many parts of the world align extremely well with the comprehensively integrated functions inherent in the MECOMS™ approach to customer service and billing.

I therefore have no hesitation in commending this book to anyone who has an interest in properly understanding and solving the many issues around managing the relationships between supplier and consumer in the complex environment that forms the modern utilities marketplace.

Roberta Bigliani, Head of Europe Middle East & Africa, IDC Energy Insights

Smart questions bring good answers.

For the utilities industry the years to come will be years of transformation. Innovation is going to permeate all activities performed by utilities.

The traditional, even unbundled, value chain will be redesigned to accommodate new business roles. Distributed Generation, Virtual Power Plant, Electric Vehicles, Storage, Smart Distribution Grids, and Prosumer Engagement, are only a few examples of originally pilot initiatives that are progressively becoming commercial realities.

Also the adoption of Mobile, Cloud, Social, Analytics and Big Data solutions is accelerating innovation and will offer utilities new means of providing business innovation, by bringing new business engagement models, new procurement strategies, different demands for IT talents, and more changes in the role of the CIO, with the CIO asked to become a major force for innovation.

In most innovative utilities, business is already demanding CIOs to co-lead innovation or at least to provide a portfolio of both technology and business services that enable rapid changes to respond to market pressures.

At the same time, business executives are becoming more comfortable and knowledgeable about ICT technologies and are increasingly taking ownership of technology-based projects, including ICT-based ones. This translates into a new perspective when considering IT investments, which are no longer viewed as "IT Investments" but are treated as any other business investment.

Flexibility or, in other terms, business agility is the "New Normal", especially for utilities operating in competitive environments.

So, where to start to successfully select and implement future proof business applications in the Utility sector? This book suggests a valuable approach: regardless your company's role, business or IT, take a holistic view and start asking yourself why, what and how. Smart questions will bring you good answers and a way of action!

Authors Introduction

Without a doubt, the energy and utilities sector is in a state of flux. To keep up with the changing environment, business applications need to be revised. This book wants to give you a helicopter view on implementing business solutions in the energy and utility sector using the MECOMS™ product on Microsoft Dynamics AX. The objective of the book is not to be a training manual on MECOMS™ or Microsoft Dynamics AX. Instead, we want to give you food for thought on how you implement smarter solutions for the energy and utilities sector.

We strongly believe in the power of the Smart Questions book series. We all are short of the vital resource "time". The Smart Questions book structures information based on the theory of Barbara Minto's pyramid. It gives you questions and hints on why these questions are important to you. It is designed to help you with the various aspects of implementing a business solution for this sector.

No, this book neither provides you with detailed insight in the energy and utilities sector nor with the business or technical details of implementing a solution.

Instead, it gives you an overall insight into important themes and provides a set of questions and details why these are important to you.

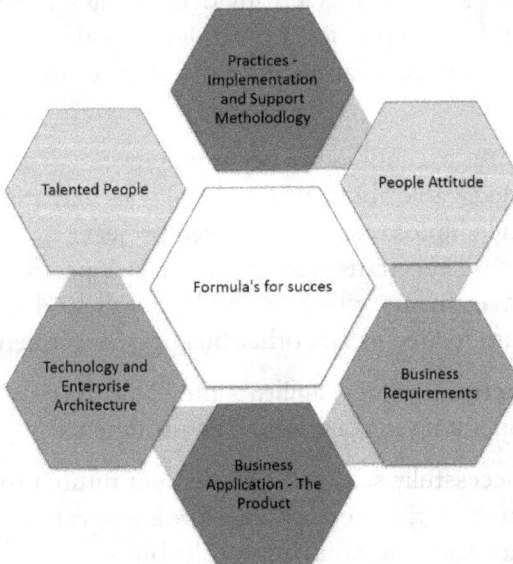

Figure 1 - The fabulous six

We strongly believe the success of an IT solution that really supports a utilities business relies on 6 important components.

We often call them the fabulous six.

- **Business Application – The Product** – This is the set of functions and features. It is about the fit between the functional requirements of the business and the product features such as the MECOMS™ product feature set. Non-functional requirements are equally important. The product/solution needs to deal with both in a cost effective and efficient way. You will need to think about how the functional, process and information models fit your business needs. You also need to develop your enterprise architecture and determine how this will evolve over time.
- **Technology and Enterprise Architecture** –A solid technology platform is vital for building robust applications. Consider aspects such as manageability, scalability, usability and ease of integration.
- **Business Requirements** – if a mission/vision/strategy exercise does not result in a clear set of business requirements, implementations cannot be successful. Success is the result of a strategy supported by the organization. This is the basis for business application requirement definition and successful solutions.
- **Talented People** – You know what to implement, how to implement it and which tools/product you will use. Is success guaranteed? No! You are forgetting the most important aspect. People! Your solution has users who need to back the implementation. By the way, the users can be your internal business users but also the utility's customers. You also need a project team to implement the solution and a support team to ensure the application maintenance.
- **People Attitude** – You can have the most talented people but if you cannot get them to work as a team, the project will fail. The human interaction aspect turns out to be one of the most daunting aspects of implementing any application.
- **Practices** - Implementation and support methodology – The implementation and the application maintenance

methodology are important to structure the work to be done. Having the right practice prevents wasting valuable time as a result of reinventing the wheel.

We have structured the book in such a way that individual chapters cover the various aspects you are facing when implementing and supporting a smarter solution for a smart utility. The key topic for each chapter is highlighted in the books table of contents below.

Chapter 1	Implementing smart solutions for smart utilities
	What makes Smart the buzz word in the utility sectors? This chapter gives and **introduction to the utility sector.**
Chapter 2	Market models for utilities
	This gives you a top level understanding of the different **market models** in the energy and utilities market.
Chapter 3	MECOMS™ solution for utilities
	The chapter gives you an overview on the **functional model** of the solution and how it extends Microsoft Dynamics AX.
Chapter 4	Utility information model
	In order to structure all the information required to run utilities processes, you need an **information model.**
Chapter 5	Utility Process model
	This chapter describes how to build the **business model and process model.**
Chapter 6	System Architecture
	Get and overview on the **enterprise architecture** of a utilities application in general and MECOMS™ specifically.

Chapter 7	Project Approach
	Learn about the **implementation and support methodology** for MECOMS™ which extends the Microsoft Dynamics AX Sure Step methodology.
Chapter 8	Data Migration
	Learn how **data migration** from existing to new systems can spoil your project.
Chapter 9	Application Management
	Your applications need to be supported and maintained. Changes occur. This chapter gives you some tips on **support and application management/maintenance.**
Chapter 10	People and organization
	The corner stone to make projects a success are the people. There is the people implementing and supporting the project and the business people using the project result. The **people** thus!

I hope you enjoy the book as much as I enjoyed writing it.

Wishing you happy, productive reading and even more importantly, thinking, planning and executing.

Dirk Michiels, the author

Who should read this book?

Why should you invest time in reading this book? We felt it was high time to write a book that gives insight in how to implement successful business applications in the utility sector.

So what is the target audience? The book aims to be a catalyst for a number of people in the organization.

People like you and me

This book is not technical nor was it ever intended to be. It is a book on how IT Solutions can support the utilities business. It offers you a bird's eye view.

It is aimed squarely at those who see IT as a utility that should be consumed to serve the business. Not the reverse. People like you and me.

Here are just a few and why it is relevant to them.

Utility Companies CEOs

The CEO takes the ultimate responsibility for strategic decisions. Choosing the right business models, processes and supporting IT Solutions is essential to the success of the organization. When looking at the options for IT Solutions in this area, the MECOMS™ solution on Microsoft Dynamics AX comes to mind.

Utility Companies CIOs

CIO's in utility companies are pulled in different directions to support the business strategy and vision. There is the need to support top and bottom line growth. Attracting and retaining customers is often vital to the business strategy. Managing and reducing cost is often a necessity. This requires having the right IT architecture in place.

Utility Companies CFOs

You need to think of Return On Investment and therefore the Cost-to-Serve a utility customer. Therefore the cost structure is top of your mind. Costs include capital and operational expenditures.

When you develop a strategy, you will need to investigate the impact of a new business solution on the IT budgets but also on operational budgets of business departments such as back office billing, metering, call center…

Utility Leadership team members, departmental heads and project team members

Anyone involved in implementing a new business application will benefit from this book. It creates a common understanding of all aspects associated with building and maintaining a new system and new processes.

Consultants and System Integrators, Project Implementers

Energy and utilities companies often rely on consultants and system integrators. This book wants to provide a common ground between what is typically called the "customer" (The utility company with its business and IT departments) and the "supplier" (The product vendor, implementation project teams, consultants).

How to use this book

This book is intended to be the catalyst for action. We hope that the ideas and examples will inspire you to act. So, do whatever you need to do to make this book useful. Use Post-it notes, write on it, rip it apart or read it quickly in one sitting. Whatever works for you. We hope this becomes your most dog-eared book.

Clever clogs – skip to the questions

We have structured the book to cover all aspects of an implementation.

Most chapters consist of a text giving you an overview of the topic followed by a list of Smart Questions.

Each Smart Question has a reason. Why does this matter to you? This approach triggers your mind to think about the choices you make.

Some of you will have a deeper understanding of the background to utility business solutions. They can skip the text and jump right to the smart questions.

Chapter

1

Implementing Smart solutions for Smart Utilities

And in knowing that you know nothing, that makes you the smartest of all.
Socrates (Philosopher, 469 BC – 399 BC)

The same old story

I toggled the switch and as always the light came on. The wind was buffeting and made it feel outrageously cold. It has been one of those days with operatic emotions. I am really looking forward to wind down. My district heating power system ensured the room had a nice and cozy temperature. I had just parked my electric car in the garage, plugged in the grid and walked inside. It felt like all the stress of the day evaporated as a light mist. I took some water from the tap, put it in a pot and turned the knob of gas stove to boil the water to prepare tea. I liked the sound of kettle whistle. Soon, I surrendered to gravity and buried myself in the sofa and enjoyed my brew whilst watching the television news headlines. It was as if it has always been like this. Nothing ever seemed to have changed. Having Water, Electricity, Gas, Heat seemed so obvious. It all looked so simple. Life is good, enjoying the luxury of the utilities. What could be difficult about all this?

People working in the sector know it is difficult. There is more to it than transporting some commodity and converting a meter reading into an invoice and collecting the cash.

Most people regard utilities as conservative, never changing organizations. Nothing changed since the light bulb and distribution grids where introduced in the late 1800's. Many people around the world take utilities services for granted. The sector is seen by many youngsters as not a cool sector to work in. It is considered a non-innovative boring sector.

Consumers do not think of the complexity of production, transport and distribution.

Figure 2 - Looking SMART

But in the recent decade, this once so static and conservative sector has been hit by a tsunami of change. Many countries have opted for deregulation, liberalization and all of sudden companies are operating in a competitive landscape. From now on winning the heart of the customers is driving the strategy of utilities.

Global warming required greenhouse gas emissions to be reduced. The race for finding new greener technologies and commercializing them was on. Commodities like water and gas are becoming scarce.

Growing populations and consumptions result in higher demands on the distribution grids. Electricity grids are also often heavily impacted by distribution production initiatives such as solar power, micro-combined-heat-power systems and wind turbines.

The energy and utility sector has changed more this last decade than it has in the entire last century.

Consumer expectations

The energy consumers, utility customers have very different characteristics than one decade ago. If they have the option, they will "zap" to a new retailer with a better offer.

New consumers are looking for added value and a holistic product experience that matches their lifestyle. This goes way beyond the traditional "take it or leave it"- utility experience. The Facebook generation has very different views on how person-to-person and person-to-company interaction should work.

Consumers want to efficiently interact with providers. Rethink, the "I am Manuel from Barcelona" call centers. Existing more traditional customer interaction channels need to be revised and expectations for new self-service, social media and other customer interaction channels needs to be explored.

Obviously this comes at a cost. Utilities must ensure they balance out the cost and the consumer expectations. Look at it as offering a smart phone and a Telco services. How do brands differentiate? Do you want and IPhone android or Windows Phone? A lot boils down to emotions.

Various utilities are offering additional products and services. This can be related to the commodity with product services like boiler maintenance services, repair services, multi-commodity services (gas, water and electricity), home energy audit, home production appliances...

A utility can also decide to offer non-energy related services such as Telco services, insurances, home access and security services, banking services or even general repair and maintenance services.

In a deregulated market, utilities need to answer the question of how to acquire and maintain customers.

Consumer will choose based on the mix of product and services the utility offers at a price he is willing to pay.

Utility companies challenges and opportunities

The utilities industry sector is impacted by change. Environmental concerns and high energy prices put pressure on the utilities strategy regardless whether they operate in a regulated or deregulated environment. Each region, each legislation has its own way of organizing and implementing a strategy for the utility sector. No country is the same. Although the underlying grid infrastructure is based on the same technologies, the legislation, geo-political and social context is different. This requires utility companies to demonstrate agility to meet the requirement of the regulatory setting, political and social-economic aspects and the new utility consumer/customer.

The market models are changing. Most countries/regions are in some mode of Deregulation. This market restructuring impacts the way in which utilities operate. The driver for Deregulation is often the economic law that competition increases benefits and reduces the price for customers. In reality, this is difficult to benchmark as the rising energy prices and the market restructuring generates a more complex landscape with more players which leads to business process and IT inefficiency. Unbundling utilities requires considerable investments in IT and business changes. Successfully changing market models very much depends on the context before the Deregulation starts. When state-owned, it is more likely that the transition runs smoothly, then when the ownership was private. Deregulation for the incumbent retailer (often the regulated, integrated utility) often means losing customers to new players in the market. Big Bang versus incremental deregulation based on region or customer type is another factor that determines the success of the transition process. More on this in the next chapter on market models.

The hunt for the customers/consumers also triggers another trend, the trend of mergers and acquisitions. The reasoning is simple. Larger utilities have more customers, make more top-line with a potential for more bottom-line. Companies are looking to achieve economies of scale but have to deal with the legacy of the past. For IT departments, mergers and acquisitions can be a nightmare. The enterprise architecture must be able to deal with this at an acceptable cost level.

Environmental concerns and the commitment to a low-carbon economy result in changing investment patterns and attitudes towards the utility sector. For example the European commission has defined a climate and energy package setting ambitious targets for 2020. These targets, known as the "20-20-20" targets, set three key objectives for 2020, a 20% reduction in EU greenhouse gas emissions from 1990 levels, raising the share of EU energy consumption produced from renewable resources to 20% and a 20% improvement in the EU's energy efficiency.

New technologies such as wind energy, solar, electric vehicles are starting to play a vital role.

In various regions demand response programs are being investigated aiming to create customer awareness of resource consumption and drive a more intelligent grid.

As in any industry, efficiency of non-core processes is also an issue. This brings to mind the topic of business process outsourcing which is often a difficult debate for established utilities. For new entrants it is a more natural choice.

Distribution grids are also impacted by the new world of utilities. Distributed production changes the energy flow in the grid. Solar power using direct current solar cells requires the installation of inverters. This causes harmonic distortion on the grid and creates challenges to maintain the voltage levels. All this new stuff needs to run on old assets. Replacing these aging assets requires a well-developed and long-term investment vision that can keep up with the smart utility vision of the future. Think of the water distribution grid and the challenges of maintaining old piping.

We should keep in mind that all this needs to be done by people. The average age of the typical utility employee is increasing. There is a definite need to rouse young people's interest in the sector.

Obviously, building a business strategy to meet the changing environment also impacts the IT strategy. The changes and their impact drive the enterprise architecture planning process. The capital and operational expenditures need to be aligned with the legislation, the market structure and all the other aspects in this paragraph. One definite change is the fact that the consumer/customer has a really central role in determining the strategy.

Resource availability

Electricity – In March 2011, the collision of two enormous tectonic plates near the coast of Japan killed thousands of people. It damaged buildings and knocked out the energy systems. This event changed the way the world looks at nuclear power and at power in general. As cascading effect the industrial production (electronic, cars) supply chains came to a halt. All this triggered rethinking of nuclear power strategies of many governments around the world. Alternative electricity sources such as gas-fired electricity plants, wind and other energy sources need to be considered when determining the long term strategy.

Water – More than a billion people worldwide lack safe drinking water. Various regions in the world are faced with a shortfall of water supplies. In the Los Angeles area the expected population growth will result in shortages. China, India, Iran, Mexico and several other countries are impacted. Rising temperatures as a result of global warming could trigger disappearance of rivers that are sourced by glaciers. Finding ways to optimize the consumption and production of drinking water is a critical and essential theme.

Gas – Natural gas is the world's fastest growing fossil fuel. The average worldwide growth of gas consumption is expected to be more than 50% over the next 2 decades. One thing is sure – the availability of natural gas is finite. Commodity prices are on the rise. Oil price has risen "times-four" over the last 5 years. Pricing depends on supply and demand and is impacted by regional conflicts. The gas price can be a real roller coaster. Forecasting and trading at the right price are essential to build and maintain the companies' bottom-line.

The earth's resources are not only used by utilities. Transport relies on fossil fuels. Every year the American people alone, drive the distance between the sun and the planet Pluto, about 3 trillion miles. The plastics industry needs fossil fuels as well.

In summary, commodities that utilities are selling regardless whether it is electricity, gas, water or thermal are all scarce resources with fluctuating prices. Utility companies need to manage the cost and selling prices of the commodity and need to take measures to reduce consumption.

Consumers need to be made aware of their consumption patterns. They must be triggered to reduce energy and resource wastages.

Smart, Smarter...

Nowadays everybody in the industry is talking about Smart. Smart Metering, Smart Grid, Smart Billing have become buzz words. However, getting a definition of all these new terms is a challenge. One thing is certain, the offering of utilities regardless the commodity they handle, is going to change dramatically.

We like to call it Smarter instead of Smart. Innovation is critical to sustainable success for any Utility. It is all about defining and implementing strategies that make more efficient use of the grid, reduce consumption and involve the consumer (or call him prosumer – as he might be producing as well!) in the end-to-end process.

Smart Meters differ from traditional meters as they record energy consumption on interval basis (15, 30 minutes, hourly,...) and communicate on a periodic (daily for example) or real-timee basis with the utility for billing and monitoring (Operational Technology, OT) purposes. This is not new. AMR or Automatic Meter Reading has delivered this function for a long time. AMR is often used in the commercial and industrial segment. The Smart Meter has some other neat features though. The communication with the utility is two-way which allows it to intervene at the consumption level and to switch load. Modern Smart Meters can also serve as gateway for other commodities. The electricity meter can connect to both the gas meter and the water meter. The electricity (Smart) meter serves as common gateway for the utility. Often the Smart Meter also interfaces with an in-home energy display to increase awareness. Initiatives are ongoing to standardize protocols and features. One of these initiatives is the Open Smart Grid Protocol. Smart meters are often seen as components for a smart grid. These are usually electricity grids that use Operational Technologies to monitor and manage consumption and consumers and suppliers production patterns to create a more reliable, sustainable and economical network.

The debate is now who will pay for the new home metering infrastructure and communication network. In a way the answer is clear, it will be the consumer. For Commercial and Industrial (C&I) customers this makes sense. For private customers this might be considered an extra cost rather than a way to deal with energy in a smart way. Legalizations and market parties need to ask themselves what the value (Benefits / Costs) of smart metering is for the consumer.

One does not need a smart meter to know that a led lamp consumes less energy than a halogen spotlight. But what is the cost of the led lamp? Will I be able to see some return on my initial investment?

To a large extend the smartness of the sector is determined by regulators and policy makers. Utilities can and must influence the policy makers. But in the end it is up to the utility companies to make sure they can build smarter processes that innovate and make them best of class. This is what we call smarter utilities. Innovation is driven be people. This is where MECOMS™ comes in handy.

... MECOMS™!

How can IT deal with this challenging environment?

IT can no longer live in a green house. When talking about Business Aligned IT Solutions and the energy and utility market in a MECOMS™ environment, we talk about:

- Customer care and billing systems often called Customer Information Systems (CIS) – This deals with interfacing with customers. The aspect is often referred to as Customer Relationship Management. Billing and cash collection is an integral part of the CIS system.
- Meter Data Management (MDM) Solutions ensure that meter data can be read using manual meter reading, self-service channels or automatic techniques and they ensure a validated consumption over a time period can be calculated. Together CIS and MDM cover the meter-to-cash cycle.
- Interaction Management – The business applications interact with customers using various communication channels such as internet, call centers, social media, voice response ... Business applications also interact with the market. Usually by a standardized market communication model to support consumptions exchange, move in, move out, switches and other market interactions. Finally, energy portfolio management is necessary in every role in the market.
- Enterprise Asset Management – The distribution grid consists of various assets. These assets all need to be installed, maintained and decommissioned. Some of the asset attributes are required to calculate conditions, consumptions, etc. An enterprise asset and service management application is required. Asset information is also vital to other roles than grid management. The retailer also needs it.

This book guides you through the various aspects that are related to implement successful solutions supporting the utility business. It shows how MECOMS™ on Microsoft Dynamics AX can be used to deal with the challenges of the changing utility industry.

The value of a business application

How do you determine the value of a solution for the business and how to compare the value offered by competitive solutions?

Making this tangible is often difficult but has never been so important. As the competiveness of the market increases, this becomes more and more essential.

Many use the key performance indicator Cost-to-Serve (One Utility Customer)to track performance. We like to look at it from the value perspective i.e. delivering superior values and getting an equitable return on the value delivered.

The figure shows a simple definition of the value. It is the ratio satisfaction over investment. Pretty obvious, the value increases with a growing satisfaction and a lower investment.

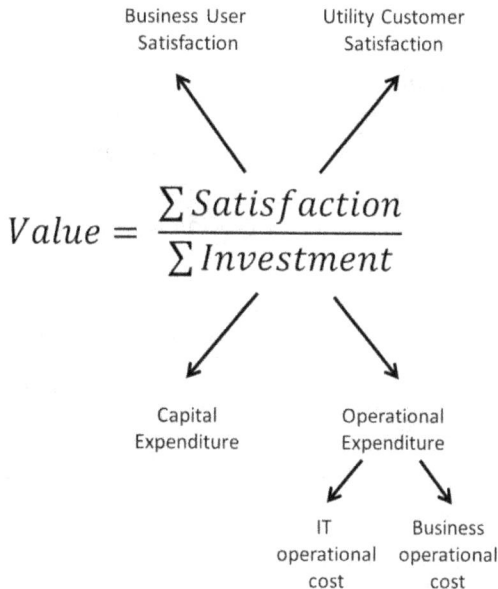

Business User
Satisfaction

Utility Customer
Satisfaction

$$Value = \frac{\sum Satisfaction}{\sum Investment}$$

Capital
Expenditure

Operational
Expenditure

IT
operational
cost

Business
operational
cost

Figure 3 - Value of utility company business application

Let's examine the terms in this equation in more detail.

- Satisfaction
  - Business User Satisfaction - The satisfaction of the business user determines the success of any business application. The usability of the application determines the efficiency of the use of the system and therefore has a positive effect on the business operational cost. You should consider the number of transactions business users can efficiently execute per day as well as the ease of access to information.
  - Utility Customer Satisfaction – Utility customers/consumers are quintessential in the equation. The values they experience determine, if legislation allows, whether they stay or "zap" away to another retailer. In the new information age utilities customers expect efficient processes and ease of access to information. Also consider the time-to-market for new offerings as a satisfaction determining factor.
- Investments
  - Capital Expenditure – The utility invests in infrastructure and business applications. It is all about managing the project portfolio of systems and processes.
  - Operational Expenditure
    - IT Operational Cost – The IT landscape needs to be managed which requires people and maintenance arrangements.
    - Business Operational Cost – What is your ratio business users and therefore business operational cost versus margin? How many people do you need to manage the business?

This way of looking at value provides a more balanced view then the cost-to-serve method.

☒	Question	Why this matters
☐	1.1.1 Is it likely that you will need to deal with new commodities?	Some utilities deal with multiple commodities like electricity, gas, water or thermal. You should consider whether this is likely to happen to your utility company as well. What is the impact on the processes? How can you create an integrated customer experience? From a customer perspective it makes sense to have one retailer dealing with all commodities. In various countries around the world, there is still a hurdle to take to go for real integrated billing. This very much depends on the political and historical context. Utilities are seen as core and essential to our lives. When comparing to the telecommunications industry where the innovation and convergence of telephone, internet and television has gone a long way, the utilities are still in their infancy. This is an area where innovation can be brought to the market. Also consider the convergence of telecommunications products with utility products.

☒	Question	Why this matters
☐	1.1.2 What is your strategy to deal with environmental concerns and involving the consumer with this?	Green initiatives are everywhere. In the Electricity world, generators play a role when choosing the most appropriate generation methods. The Distribution Grid operator needs to develop a "green-grid" that allows support distributed local production using wind, combined heat power, solar, etc. Real impacting green initiatives are the ones that impact consumption. They require the buy-in of consumers. The consumer needs to be informed and feel involved with the initiatives. This requires education and guidance for this business transition but also ensuring that processes are implemented that keep consumers informed.
☐	1.1.3 How do you intend to bring smartness into the business?	When utilities think about bringing smartness into the business they think about launching innovative initiatives and are looking for "the Idea" that will change their business. If you want, there is a Utility focused seminar or fair every day somewhere in the world where you can get a lot of ideas and inspiration. Getting an idea is only the beginning. A journey starts of developing and testing it for viability. You will also need to build a team around it, a team that drives the innovation. The Business Model Canvas techniques highlighted in the process model chapter can help you to structure and drive innovation efforts. Think of the impact innovation has on the underlying IT infrastructure and how you will be able to model the changes that innovation brings in the enterprise architecture.

☒	Question	Why this matters
☐	1.1.4 Is it likely that you will need to deal with new commodities?	Some utilities deal with multiple commodities like electricity, gas, water or thermal. You should consider whether this is likely to happen to your utility company as well. What is the impact on the processes? How can you create an integrated customer experience?

From a customer perspective it makes sense to have one retailer dealing with all commodities. In various countries around the world, there is still a hurdle to take to go for real integrated billing. This very much depends on the political and historical context.

Utilities are seen as core and essential to our lives. When comparing to the telecommunications industry where the innovation and convergence of telephone, internet and television has gone a long way, the utilities are still in their infancy.
This is an area where innovation can be brought to the market.

Also consider the convergence of telecommunications products with utility products. |

Chapter

2

Market models for utilities

People can have the Model T in any color--so long as it's black.

Henry Ford (Car Manufacturer, 1863 - 1947)

No Utility is the same

SO, how does the Utility market work? If you are a utility veteran, you might want to skip this chapter. We have included this chapter for people with limited insight in the utilities market.

Utility companies in different countries around the world use very different business models. Although the service that is delivered to the consumer seems to be very similar, there are massive differences in business models and processes. Factors that influence the business model are:

Figure 4 - Connecting in the market and IT landscape

Resource Availability

The strategy to determine the energy resource used to create energy or other commodities for your consumer will largely depend on availability. Non-renewable- (coal, gas, oil, etc.) or renewable- (sun, sea, wind, biomass) resources can be used. In the water market, it is about collecting sufficient quantities of high quality water requiring minimal processing to make good drinking water throughout the year.

Transport Grid availability

In the gas market, availability of gas depends on transport pipeline systems or ships. This determines the availability of energy resources to be sold to consumers. Depending on the geography and political context it is also possible to transport commodities across country borders

Electricity production companies and suppliers/retailers are sometimes faced with insufficient transport capacity to meet their contractual commitments. An example is Switzerland which is assumed to be the European power house. The Alps make the conditions ideal for harvesting hydropower in its more than 500 hydropower plants. This accounts for more than 50% of the country needs. If market parties want to sell this energy in other countries, the question is: "is the transport capacity available?"

Political and legislation

Many utilities worldwide evolved from services initially offered by the government or public sector. Energy and water availability, price and environmental impact are high on the agenda of governments and legislators worldwide. Initiatives such as the Kyoto protocol are efforts to come to some global alignment of policies and strategies. Hence, this puts pressure on the market regardless whether they operate in a regulated or deregulated market. Public utilities are mostly natural monopolies. The reasoning is that the investment in infrastructure is sky high. Open access to the grid allows other parties to join the market and create liberalization. Hopefully, deregulation leads to a competitive landscape creating benefits for the consumers.

Social situations

Success in the sector, largely depends on consumer behavior and the public perception of energy/utility technologies. Understanding consumer behavior determines the success of demand side management projects.

Several countries subsidize energy pricing either by giving energy or water completely for free or by giving a certain volume for free. Social schemes often determine the way people who are not paying their utility bill, continue to get access to the grid. This all impacts the way the market works.

In some countries some commodities are considered by its inhabitants as free of charge. An awareness scheme, taking into account the possibilities on the field can change this attitude.

Commodities

Utilities deal with various commodities that each have specific properties and challenges,

Electricity

Electricity is mostly sold as Kilowatt hour. Kilowatt-hour (kWh) is a unit of energy. Consumers know it as a billing unit for energy delivered to their homes. If you have an electric appliance rated at 1000 watts or 1 kilowatt, it consumes 1 kWh when used for one hour.

AC (Alternating Current) cannot be stored. If you want to store electricity you need batteries. If AC power is a requirement you will need to convert the batteries DC power to AC power.

Gas

Gas can be sold using various units of measure. One could use Cubic Meter (or feet) (M3), Kilowatt Hours (KwH) or specific measurement such a British thermal Units (BTU).

Gas is compressible and its energetic content not only depends on the temperature/pressure but also on the calorific value of the gas. The calorific value is a measure of energy contained in the gas. Moisture in the gas affects performance which requires the calorific value to be averaged and measured for a consuming point.

An important property of gas is the ability to be stored in tanks or underground. Often the liquefied state is used.

Water / Sewage

Water is billed in liter or cubic meters. Bacterial, chemical pollution and safe transportation and distribution are obvious challenges for this commodity.

Thermal

District heating and cooling systems deliver energy from a central facility. Facilities can support Industrial / Commercial and Residential consumers. Unit of measures include Kilo Watts Hour or Joule.

Roles in Market

Various roles can be identified in the market. We have included some simplified and somewhat generic architecture figures to describe the physical grid for electricity, gas and water.

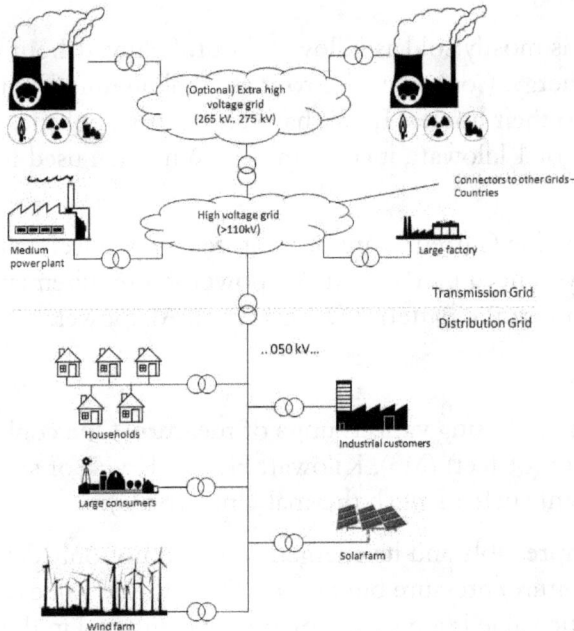

Figure 5 - Electricity grid and market parties

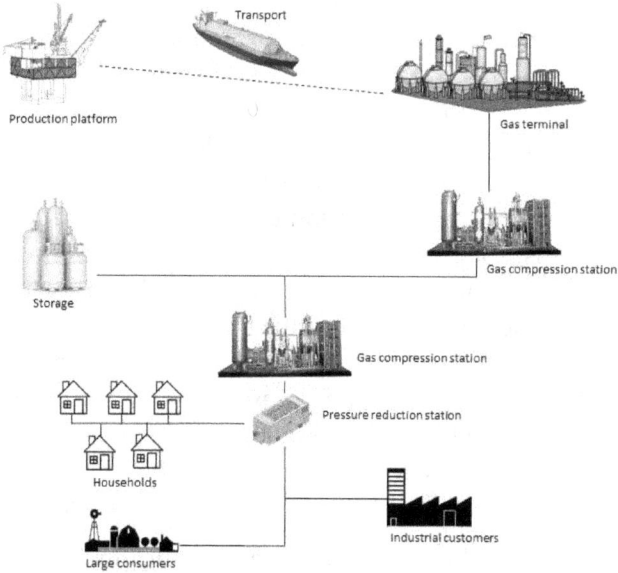

Figure 6 - Gas grid and market parties

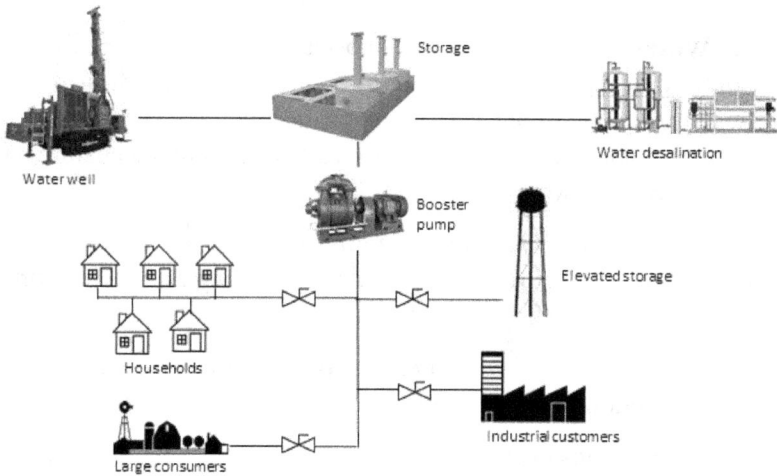

Figure 7 - Water grid and market parties

Generation Company

Generation (production) companies create the commodity.

For **Electricity** this can be done by driving an alternator/turbine with steam generated from a nuclear reaction or burning fossil fuel such as coal, oil or natural gas. In addition renewable resources such as water (using a turbine), wind turbines, biomass or photo electric solar cells can be used. Historically large central power production facilities were built. Now there is a trend in the market to go for decentralized production using Combined Heat Power (CHP) systems, solar panels and wind turbines. The consumer can now be an energy producer by investing in CHP or solar systems and injecting energy in the grid. This trend has a tremendous impact on energy flows in the grid. For example neighborhoods with a high density of solar panels can start to behave as a production facility rather than as energy consumers. DC energy generated by solar cells needs to be converted to AC power using inverters which creates harmonic distortion in the grid.

In the **Gas** world, the story starts at the wellhead where natural gas is flushed out of the ground. The gas is treated to purify it and to remove water, petroleum liquids and other unwanted elements. For ease of transport and storage, gas can be liquefied to LNG (Liquefied Natural Gas) at very low temperatures. LNG only takes 1/600th of the volume of the gaseous state of Methane (natural) gas.

In the **Water** world, production is about collecting water either through pumping groundwater or surface water. Regardless of of the way the water is collected, a purification plant is required to remove biological, chemical and physical contaminants ensuring a constant drinking water quality.

For the **Thermal** world, most systems are based on heating or cooling water in a central facility and transporting and distributing it to the consumers.

The business model of the generator/producer is based on making a profit on selling the commodity. Fuel prices are volatile.

The story becomes even more complicated as the sector is heavily influenced by CO_2 emissions trading.

This is an area where deregulation is very likely. And by having various competitive production companies in a market there is a constant price pressure.

Transport Grid Operators

Transport Grid Operators, TGO, also called Transmission System Operators (TSO) are responsible for transporting electrical power or gas on a regional, national or even cross-country border level. The cost of implementing the transmission grid for gas and electricity is very high. This makes TSO mostly natural monopolies and they therefore operate under legislator determined policies and legislations.

Electricity is transmitted (usually with overhead lines) at high voltage levels (110KV and more) to reduce energy losses due to long distance lines. Gas is transported at high pressure, for efficiency reasons,.

Distribution Grid Operators

Distribution is the final stage of delivering the commodity to the consumer.

For gas, the pressure of the transport network is reduced to low consumer pressures. For electricity, the high voltage of the transport grid is transformed to middle voltage (50 KV ranges) and distributed on the distribution grid. Near the consumption location (Commercial & Industrial Location) or street level the voltage level is brought down again to the familiar 130 or 230 Volts. More and more distribution companies need to deal with injections of energy on the mid voltage level by smaller production units such as wind and solar farms.

Like Transport Grid Operators, Distribution Grid Operators are a natural monopoly. It does not make sense to have more than one grid.

Metering Companies

A function is required to read the meters and treat the meter reading results by calculating the consumption and validating the plausibility before forwarding them to market parties that require them. This function can be operated in a competitive environment.

One also needs to consider who the owner of the meter is. This can result in more market roles such as meter asset managers.

Balance Responsible and Shipper (gas)

This is the role that ensures the balance between the supply of energy and the anticipated consumption of energy during a given period. The role is also responsible for financially settling and regulating any imbalances between supply and demand based on contracts and actual consumptions.

Supplier/Retailer

The supplier or retailer is the interface for the consumer. This is a role that can easily be deregulated.

Energy market / Spot Market

The Energy Spot market allows parties that have an imbalance between the energy they have contracted and the actual consumption or producers, to locate available buyers and trade the energy at a negotiated price.

Clearing House

In a deregulated market an organization is required that independently maintains the relationship with market parties and the consumer. There needs to be one single version of the truth to establish who the supplier of a customer is.

Regulated Market

Providing energy using a distribution grid started becoming mainstream in the late 1800's when facilities, producing electricity for their own use, supplied excess power to other parties. Governments, tramways and public lighting were amongst the first customers. Early 20th century, sales of electric appliance such as heaters, motors… took off and with it there was an explosion of electricity generation and distribution. Supplying the electricity to domestic or commercial and industrial consumers became the responsibility of public utilities. The public utilities were either a public authority owned by the governmental organization or municipalities or a private organization that was given the task to do this work as regulated monopoly.

Nothing substantial changed till the 1990's. The utilities were dealing with all the processes. Getting a new connection to the grid and disconnecting was the responsibility of the same organization that provides you with an invoice. There was no free choice of supply.

Regulated environments are still very much the norm in the water industry and the thermal industry. Much of the gas and electricity market has gone in deregulation or liberalization mode.

Deregulated market

People started wondering how to break the monopolies and felt that introducing competition would have a positive impact on customer service and that the price of the energy would go down. The market had to evolve. Utilities needed to change their rather administrative and technical patterns to become commercial and competitive market players, dealing with risk management.

Deregulation or liberalization can be implemented in various ways. No wonder that individual countries have introduced it in different ways. Even in large trade zones, such as the EU, where directives were imposed in 1997, the implementation very much depends on local country implementations.

Not a lot sympathy can be scored in liberalizing the distribution and transport grids. Therefore, they are often called natural monopolies. It does not make sense to have competitive grids supplying consumers.

For generation (production) this is a different story. Various generators can offer energy that has been produced in different ways. This creates a competitive landscape for the energy commodity.

The retail part can also be liberalized. Retailers/suppliers compete for the consumers based on price and value offered.

A clearing house or access register is required to keep track of who is supplying an individual consumer.

Supporting functions, like meter reading and meter asset management, can also be deregulated.

The success of liberalization is often not obvious. It requires balancing between the price and market transparency with consumer protection and market players incentives to innovate and invest in the market whilst ensuring security of supply and dealing with the environmental concerns.

Franchising Market

Sometimes it is thought that the overall success can be achieved by setting up franchise agreements. The franchisor is often the regulator or the government who feels that it is better to draw on the practices and business models of the franchisee than to manage the utility itself. The franchisee has a greater drive in the business as he has a direct stake in the business.

This model has been successfully used in many fast food companies such as Subway and McDonald's.

As with deregulation clear rules need to be defined to play the game. Franchisees are usually contracted for a long period. Solid service level agreements must be set up to monitor and manage the performance of the franchisee.

Implementing a franchise model is often restricted by the social passive of the existing utility. It requires good balancing to create a win-win situation for the franchisee and franchisor.

☒	Question	Why this matters
☐	2.1.1 What is the role you are playing in the market?	This seams a trivial question. But with the changing environment in utility land, it is essential to think about likely evolutions in your role in the market. This might require adding some competitive environment thinking.
☐	2.1.2 Is there a well-defined plan on what your winning strategy is for this market?	Each role in the market must think of how it can win in the market. What is the long term objective of the organization? Kaplan Strategy map thinking will help with this. We like to use the Business Model canvas model method outlined in the process chapter to understand how the strategy is defined.
☐	2.1.3 Is the value proposition to the customer clear?	Defining a value proposition to the customer is a carefully designed equilibrium between revenue and cost generating aspects. The value proposition is what will attract customers to use the products and services of the company. Many utilities operating in a monopoly do not understand the concept of customer. They think of connections where their commodity is delivered. Rest assured, regulators worldwide are changing this mindset.

Chapter

3

MECOMS™ Solution for Utilities

The greatest challenge to any thinker is stating the problem in a way that will allow a solution.

Bertrand Russell (British Mathematician & philosopher, 1872 – 1970)

What is MECOMS™?

MECOMS™ is an integrated solution for Energy and Utility companies covering their core processes Customer Care and billing, Customer Information System, Meter Data Management as well as their supporting processes.

MECOMS™ builds on the Microsoft Flagship ERP Solution, Microsoft Dynamics AX. This allows Ferranti, the developer of MECOMS™ to focus on the utilities functionalities whilst the multi-billion dollar development machine of Microsoft focusses on getting the baseline technologies and applications right. These technologies range from

Figure 8 - The uprising of flexible - smart solutions for utilities

user interface, over portal, database and reporting to office integration and Business Intelligence. The Microsoft Dynamics AX platform also offers robust business solutions, ERP functionalities and processes. These functionalities include:

- Finance (General Ledger, Account Receivable, Account Payable, cost accounting, bank interface, budgeting)
- Project Administration
- Inventory Management
- HR (recruitment, talent management)
- Service Management
- Shop floor control / production

MECOMS™ Solution components

Microsoft Dynamics AX

MECOMS™ extends the Microsoft Dynamics AX application. The architecture of Microsoft Dynamics AX is designed to support market verticals. The market verticals use the core Microsoft Dynamics AX functionalities and extend the information model, the functional model and the process model (more on this further on in this chapter and in the next two chapters) to support the typical vertical requirements.

MECOMS™ is a vertical for Microsoft Dynamics AX that supports small, medium and large enterprise requirements of the energy and utilities market. As a result the user has access to all the ERP functionalities and the MECOMS™ vertical utilities functionality and processes.

Core

MECOMS™ EUCA, The Energy, Utilities Community Architecture is the core of MECOMS™. It contains the basic data structure and logic to support the Energy and Utility segment. Due to the object oriented nature of Microsoft Dynamics AX, MECOMS™ EUCA extends the ERP objects to deal with the Energy and Utilities sector information requirement.

MECOMS™ EUCA is the informational model that allows you to support the complexity of the energy and utilities industry such as metering configurations, metering registers, consumptions, contracts, service requests, invoices, invoice lines and much more. At first glance they look pretty generic but if you work in the utility sector you know that they are very complex.

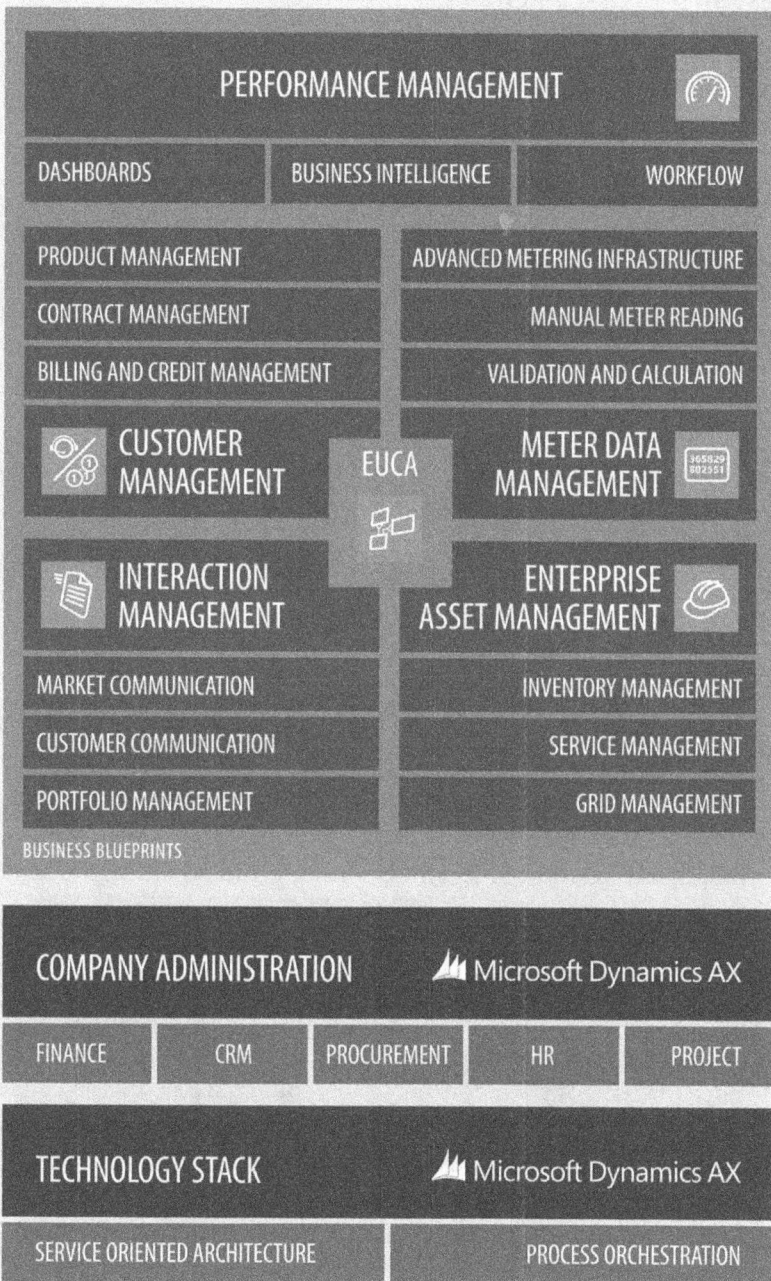

PERFORMANCE MANAGEMENT

| DASHBOARDS | BUSINESS INTELLIGENCE | WORKFLOW |

PRODUCT MANAGEMENT	ADVANCED METERING INFRASTRUCTURE
CONTRACT MANAGEMENT	MANUAL METER READING
BILLING AND CREDIT MANAGEMENT	VALIDATION AND CALCULATION

CUSTOMER MANAGEMENT — EUCA — **METER DATA MANAGEMENT**

INTERACTION MANAGEMENT — **ENTERPRISE ASSET MANAGEMENT**

MARKET COMMUNICATION	INVENTORY MANAGEMENT
CUSTOMER COMMUNICATION	SERVICE MANAGEMENT
PORTFOLIO MANAGEMENT	GRID MANAGEMENT

BUSINESS BLUEPRINTS

COMPANY ADMINISTRATION — Microsoft Dynamics AX

| FINANCE | CRM | PROCUREMENT | HR | PROJECT |

TECHNOLOGY STACK — Microsoft Dynamics AX

| SERVICE ORIENTED ARCHITECTURE | PROCESS ORCHESTRATION |

Figure 9 - The MECOMS™ Functional Model

MECOMS™ Business Domains

Customer Management

The Customer Management business domain deals with all customer related aspects and includes following functionalities or subdomains:

- Product Management - MECOMS™ provides unparalleled flexibility for managing products in today's increasingly competitive utilities markets. Users can create and maintain products by linking commodities with rate structures, taxes and levies. A product configuration contains all parameters for billable items and services. MECOMS™ enables utilities to rapidly develop new market offerings, without any additional programming.

- Contract Management - Contract types and templates are created according to the products offered. MECOMS™ features an advanced pricing and quotation suite which is especially valuable in today's B2B markets. It allows fixed, semi-flexible and full flexible pricing models. The formulas to calculate prices are managed using an intuitive, graphical formula builder.
  Advanced contract conditions, such as auto-balancing, cash-out and take or pay, are all possible. MECOMS™ automatically takes care of these arrangements in billing and other administrative processes.

- Billing and Credit Management - Billing and credit management are core processes for any utility. MECOMS™ is designed to rapidly deliver impeccable bills, to encourage timely payments and to maximize the payment rate of every customer segment. MECOMS™ lets utilities maximize the amount of timely payments, while minimizing the dunning costs.
  The specifics of each contract and product are seamlessly taken into account by the billing system. It calculates and validates invoice-lines on-the-fly as meter and consumption data enters the system. This workload spreading allows for an amazingly fast billing run when the actual invoices need to be generated. Furthermore, on-the-spot validations detect anomalies long before the actual billing run. This allows error-free billing runs, with increased speed and accuracy. Furthermore, MECOMS™

provides out-of-the-box support for real-time pricing and billing. Scheduled and on-demand meter readings are handled simultaneously and future price changes are already taken into account when calculating advances.

Rules for validation are easy to configure and provide flexibility to cope with unexpected events. Besides metered commodities, a system of consumption units allows billing of unmetered product.

Furthermore, additional items can be derived from the metered consumption. For example, calculating waste water charges based on tap water usage.

MECOMS™ Credit Management handles payment processing and all communications related to it. Bills can be presented in various ways: through a web portal, per e-mail, by post (using a print service) or through standard interfaces, such as EDI.

MECOMS™ assigns and regularly updates scores to assess the creditworthiness of every customer. Furthermore, information from external credit checks can easily be linked to the customer database to refine the risk assessments. This is especially useful for B2B customers.

A sophisticated engine for collection management or dunning, takes into account the customer's segment and individual history. It then determines whether to charge a fee, after which it will send out or plan the most appropriate reminders (letter, e-mail, personal call). If this has no avail, MECOMS™ provides methods to send defaulters to a collections agency or to initiate legal action.

MECOMS™ offers flexibility to adjust payment terms to the specific situation of a customer. Measures such as delaying due dates, custom payment schedules and payment by third parties enable the customer to settle his debt instead of defaulting.

MECOMS™ communicates directly with banks and automatically interprets all incoming payments to manage direct debit administration. At the core is a sophisticated matching engine. It combines many parameters into a dynamic scoring system to decide which payments can be processed automatically. Only when there is too much uncertainty, the system will request human intervention. It will then assist the user to decide by listing the different possibilities based on their probability.

Meter Data Management

- Advanced Metering Infrastructure - MECOMS™ is smart meter ready. It provides two-way communication with smart meters and is able to process large volumes of smart metering data via the Smart Metering Communication Bus. Moreover, MECOMS™ facilitates smart metering processes for meter asset management, monitoring and the interaction with customers and other market parties.

- MECOMS™ also handles Automatic Meter Reading (AMR) as a core process. It can easily integrate with existing systems, such as Itron, Echelon/NES but also communicates natively with other automatic meters, without additional appliances. An efficient unmetered to metered flow facilitates large-scale meter deployments and offers bulk transferring connections from specific areas.

  The MECOMS™ Smart Metering Communication Bus (SMCB) has been designed to manage Smart Metering assets and complex two-way communication in a multi-brand, multi-protocol and multi-user environment. SMCB guarantees a comprehensive management of the metering assets, while at the same time offering the flexibility to link several different proprietary protocols to each other and to the central system. In addition, it supports "push" as well as "pull" AMI architectures. SMCB makes MECOMS™ a device-independent solution.

  SMCB is developed together with database experts from Microsoft. MECOMS™ uses innovative technologies, such as In-Memory Analytics and Smart Volume Streaming, to ensure that it is able to handle massive amounts of data. In a 2012 benchmark, a few regular PC's running MECOMS™ were able to handle 270,000 metering values per second. In a real life situation, with 15 minute intervals, this would correspond to more than 100 million meters.

- Manual Meter Reading - MECOMS™ organizes the entire Manual Meter Reading process. MECOMS™ wireless portable terminals for meter readers enable market parties to receive their meter readings within the same day. MECOMS™ Mobile lets the meter reader staff access and register relevant information on their terminal (e.g. defects, fraud or the meter location). The terminal performs on-the-spot validation and can communicate immediately using a mobile data connection. It also provides meter readers with their assignments for the upcoming days.

Periodical meter reading routes often reflect the accumulated experience of many years. For unplanned meter readings on demand (MROD), however, routes are generated dynamically. MECOMS™ takes into account the location of the meter reader and the accessibility of the locations to create an optimal route. The continuous connectivity of the terminals allows operators to send updates or additional tasks to a meter reader while he or she is in the field.

MECOMS™ also offers many self-service channels. Paper index cards returned by customers are automatically scanned using optical character recognition (OCR). An interactive voice recognition (IVR) system lets customers report their meter readings over the telephone and can include immediate validation. Furthermore, MECOMS™ B2C and B2B web portals let customers enter index values, perform validations and consult their consumption history.

- Validation and Calculation - Meter readings are stored and validated in MECOMS™ using flexible business rules. Even on ordinary hardware, MECOMS™ smoothly processes and stores millions of meter readings. MECOMS™ allows the definition of business validation rules in a user-friendly manner, without additional development. Multiple rules can be active simultaneously and can be prioritized to match changing business needs. The validation rules are executed in the main MECOMS™ system but also on mobile devices and self-service portals to further reduce the likelihood of erroneous values. Converting meter readings into consumptions is done using formulas in meter certificates which are linked to the meter of each individual customer. Thus, readings of a diverse, historically grown metering park are processed consistently. Complex metering configurations, often with local production, are also taken into account. MECOMS™ makes it easy to configure calculation factors and caloric values, such as daily temperatures (to calculate energy in gas volumes), SLP (synthetic load profiles) /EAV's (estimated annual volumes) and proxy-consumptions (related products which are metered indirectly). Estimations, based on consumption profiles or historical data, are automatically calculated as a substitute for missing data.

Interaction Management

- Market Communication - IT systems do not exist in a void. Interfacing with different systems is required to exchange messages, to update information and to account for all business processes in a multi-player market. This is the role of market communication, both in a liberalized and a non-liberalized market.
Market Communication enables SOA-based EDI or XML messaging of customer scenarios, master data and meter readings and invoice information. MECOMS™ can also take care of settlement processes, by performing grid fee calculation, allocation and reconciliation. In a regulated market, Market Communication can be used for reporting to partners, regulators, etc.
The MECOMS™ design separates generic processes from market-specific business logic, using interchangeable market blueprints. This approach greatly simplifies new implementations and continuing compliance with evolving market rules. MECOMS™ also provides a standard message library and a message mapper tool to rapidly implement new messages. Journals allow easy follow up and error-handling.

- Customer Communication – Advanced CRM – As discussed previously, customer and user satisfaction are of utmost importance in today's utilities industry. MECOMS™ enables utility companies to improve customer service while reducing Cost-to-Serve at the same time. A high degree of automation, intuitive user interfaces and sophisticated error prevention increase both efficiency and customer satisfaction. Customer care staff is empowered to handle even the most complex situations in fluent workflows, while dynamic rules make it easy to customize MECOMS™ for specific situations.
In order to increase customer intimacy, MECOMS™ offers innovative interaction channels. B2B and B2C web portals let customers consult and update their information with a web browser. Furthermore, for B2C customers, MECOMS™ offers a smart home application which runs on Xbox game consoles and an App for tablets.
MECOMS™ CRM offers customer care staff a 360° Cockpit™

for a quick and complete view on all information related to a customer, from meter readings and consumptions, to contracts, invoices, payments, repair and maintenance. State-of-the-art screens enable real-time completion of complex processes, such as move-in, final invoice, contract creation and catch-up of payment advances. With the 360° Cockpit™, processes are completed so fast that the customer can remain on the phone until completion. This significantly reduces call times and improves the experience, both for customers and employees.

- Portfolio Management - MECOMS™ Portfolio Management offers a comprehensive toolset for program management, portfolio analysis, position calculation, risk management and mark-to-market. Users can set-up and develop a forecasting model using a deal/trade subsystem, a configurable calendar and a formula generator.

  Based on the results, companies can manage their risk and position within the market. This includes sourcing activities (buy/trade) on spot and futures markets, often using flexible contracts in the MECOMS™ contract management module.

Enterprise Asset Management

- Inventory Management - Utilities need to manage thousands or millions of meters and other devices, often at customer premises. These need to be installed, maintained, calibrated, repaired or replaced. MECOMS™ Inventory Management tracks all assets, from pipes and cables to connections and stores communication properties for Automatic and Smart Metering. Furthermore, Asset Life cycle Management manages the complex and evolving relations between assets and provides a way to go back in time and trace each asset throughout its lifespan.

  When a technician works on an asset, he can access all related information from a single screen. Furthermore, any information provided by the technician about the meter's condition is fed into the Conditional Monitoring module. This will allow the creation of automated actions based on pre-defined situations.

- Service Management - MECOMS™ Service Management synthesizes all available information to optimize both planned and incidental maintenance. It plans and assigns maintenance work to field workers, for example meter installations and decommissioning. Using mobile devices, technicians work on-site while being connected in real-time to the back office system via GPRS. Rescheduling, communication, alerts and notifications are all done using these handheld devices. Information about skills and people's availability from the HR module is used to propose an optimized planning and dispatching of resources. Furthermore, MECOMS™ uses project and quotation templates to improve financial transparency. Work outsourcing or offering added services, such as service contracts for devices owned by customers, are seamlessly integrated into the overall service management. By eliminating inefficiencies and unnecessary work, productivity is significantly improved.

- Grid Management - MECOMS™ Grid management lets utilities define, consult and navigate delivery and data communications in a time sliced view. To avoid needless incident reports, maintenance and shutdowns are automatically communicated to the validation process and to the CRM staff while they take calls. Furthermore, Grid Management can offer visual GIS-integration, smart metering provisioning using service pools and tight integration with the smart metering communication bus.

MECOMS™ Performance Management

MECOMS™ Performance Management contains sophisticated tools to analyse data and provides innovative ways to make the insights available to the whole organization.

Dashboards

- MECOMS™ offers pre-defined role centres for common job functions in utility organizations. These offer an intuitive approach to use the system and help to cope with the natural overlap between different modules and departments. These web-based screens provide a unified view with all relevant work cues, navigation, links, KPI's and graphics, so managers can quickly asses their unit's performance. Apart from common financial and operational KPI's, any workflow can be turned into a KPI, to assess its speed and outcome.

Business Intelligence

- MECOMS™ lets utilities measure their effectiveness by transforming data into intelligence. Profit Management, based on customer segmentations, margin calculation and Cost-to-Serve analysis, empowers energy and utility companies to identify businesses opportunities and take informed decisions. Furthermore, users can configure their own KPI's, such as environmental impact or staff productivity. Past trends can be analysed to forecast future evolutions and data can easily be exported to other analytical tools.

Workflow

- MECOMS™ Workflow is a generic framework to automate the steps of any process. It greatly improves cooperation between departments and increases accountability. Furthermore, the status and results of any workflow can easily be included as KPI's in a role centre.

☒	Question	Why this matters
☐	3.1.1 What core general business processes are available in the product?	The MECOMS™ solution is built on the flagship Microsoft Dynamics AX solution. As a result the rich set of functionalities for Finance (GL. AR, AP, Budgeting, etc.), Human Resource management (recruitment projects, training management, skills management, appraisal management, etc.), project and much more are available. The integration capabilities also allow MECOMS™ to be integrated with existing ERP solutions.
☐	3.1.2 How much bespoke development is required to meet your functional requirements?	The advantage in using a standard common off-the shelf solution such as MECOMS™ is the richness of the out-of-the-box functionalities. The software still requires some configuration tailored to specific uses. By using out-of-the-box functionalities of the Microsoft Global ISV and Microsoft certified for Dynamics AX solution the MECOMS™ product reduces the overall system-development costs and the application maintenance and management costs. By reducing the cost – to – serve per customer we therefore increase the value of the solution. Develop a functional fit gap list to get a feel of the overall functional fit.
☐	3.1.3 Which underlying data model is used for each of the functionalities?	The MECOMS™ EUCA information model is the common data store for utility data. MECOMS™ EUCA extends the core Microsoft Dynamics AX ERP data model. There is only one (1) information and data model for all utility specific and general ERP functionality.

☒	Question	Why this matters
☐	3.1.4 What is the level of completeness and fit of the Meter Data Management functionalities in terms of meter reading?	*In terms of "read" functionality of MDM.* MECOMS™ offers the flexibility of connecting to various solutions. It can deal with advanced metering infrastructure head end systems connecting to the meters in the field and/or use manual meter reading functionality and/or self-service (using interactive voice response, self-service web portal,…) and market provided meter reads.
☐	3.1.5 What is the completeness and fit of the Meter Data Management functionalities converting meter reading in validated consumptions?	*In terms of "treat" functionality of measurement data. Or call it VEE (Validation, Estimate, Edit) and calculate:* Think of how the solution manages the specifics of the commodity: • Can it handle converting factors introduced by voltage and current transformers to convert the meter reading in consumption? • Can it deal with caloric values, temperature, etc. to convert a gas reading in consumption? What is the functionality for plausibility check? • Which are the available validation rules? Do they work on meter reading and/or consumptions? Can it use Synthetic Load profiles to deal with estimations? We call this the "treat" functionality. This is a necessity to calculate the consumption from the meter read.

☒	Question	Why this matters
☐	3.1.6 Can the solution deal with the high volumes of meter data readings and make them available to the applications that need them?	*In terms of "read" and "store" functionality of MDM.* Can the solution deal with the information avalanche associated with the (smart) meter data management function? MECOMS™ uses the Microsoft Big Data strategy to deal with this. MECOMS™ uses HDInsight which is Microsoft's 100% Apache compatible Hadoop distribution, supported by Microsoft. HDInsight, available both on Windows Server or as an Windows Azure service can be used by MECOMS™ to store large volumes of information in a very cost efficient way.
		Think about the volume of data you need to handle.
		Or in general, how complete is the solution to cover the *Read-Treat- Store* cycle of meter readings for the volumes it has to deal with? Create a fit/gap list to understand the match of the MECOMS™ solution with your requirements.
		The meter readings are required to make various calculations such as converting the meter readings to consumption, energy auditing and much more.
		Meter Data Reading Storages is an integral part of the MECOMS™ solution. It stores all this information in one integrated solution. From there the information can be distributed over various systems.

| ☐ 3.1.7 What is the level of completeness and fit of customer management functionalities converting consumptions in cash? | Customer management needs to be able to define the product you are offering in the market. Products have associated tariff structures which allow to create sales proposals and contracts offered to customers, in a structured way. The MECOMS™ customer management module offers this functionality out of the box including the functionality to deal with market price fluctuations. Based on the contract, the consumption and the details of additional provided services invoices are generated by a mass billing process. You need to consider the volumes for the billing process. Think of a utility with 2 Million customers and, as a result of the contract structure, about 10 invoice lines per invoice. This creates 20 Million invoice lines per month. You need to consider how this process runs and how errors in the process are handled. We cannot stop the process for all customers if one customer invoice cannot be generated for whatever reason. The solution needs to be able to support a plausibility check on invoices. You obviously make invoices to collect cash. Larger utilities cannot do the matching of invoices manually. An automatic payment matching process needs to be in place. Usually this also requires an interface to direct debit services and cash collection agencies. A quick note – Your billing process is not always a customer mass billing one. Think of distribution cost billing; energy purchase, etc. Create a fit/gap list to understand the match of the MECOMS™ solution with your requirements. |

☒	Question	Why this matters
☐	3.1.8 What is the level of completeness and fit of interaction management functionalities dealing with customer and market interaction channels?	Utilities do not operate in isolation. They need to exchange information with external parties. In a deregulated context this includes various market messages to make the de-regularized market work. The market interaction messages typically depend on the market and market role and could include messages for: • Customer Switch • Supplier Switch • Move in / Out • Meter Data Readings • Etc etc... Depending on the regulation, the information can be supplied using EDI, XML, CSV formats. Typically a work flow deals with the business processes that result from receiving messages. The Market Communication module of MECOMS™ is a configurable framework that deals with the complexity of interacting with the market.

☒	Question	Why this matters
☐	3.1.9 What is the level of completeness and fit of the Enterprise Asset Management functionalities?	Utilities need to deal with a whole set of assets. For the meter-to-cash process having the exact information and attributes of the meter is essential. Meters are installed and replaced. This process is supported by the MECOMS™ Enterprise Asset Management module. Enterprise Asset Management works fully integrated with the customer communication module of Interaction Management. Customer calls requiring a service ticket to be opened can be handled by the Service Management functionality. The solution maintains the integrity of the Asset database by using the functionality of the inventory management module. This functionality not only applies to meters but to any asset in the utility.
☐	3.1.10 What is the level of completeness and fit of Business Intelligence functionalities that allow you to monitor the business performance?	Often utilities spend time on structuring processes but forget to think about process performance. This is what performance management is all about. MECOMS™ uses the SQL Server Analysis and reporting services that come out-of-the-box with Microsoft Dynamics AX to offer these functionalities in the user's role center. Microsoft Power Pivot and Microsoft Power View are essential technologies to offer ad hoc Business Intelligence to the users. Microsoft is regarded as the top player in this market space.

Chapter

4

Utility information model

It is a very sad thing that nowadays there is so little useless information

Oscar Wilde (Irish Dramatist, 1854 – 1900)

Why is this important?

I N this chapter, we want to look at individual data elements that are required in a Utility solution. We will also relate each of the elements to the physical world.

The MECOMS™ Energy and Utilities Community Architecture (EUCA) defines the information model dealing with utility business applications. It is not our intention to give you a full insight in the EUCA model given limited space and time available. Instead we want to familiarize you with some core concepts.

The information model required to deal with the complexity of meter data management and customer information systems

Figure 10 - Structuring the information model

exceeds by far the information model of standard ERP and CRM solutions. The information model is the heart of the solution but at the same time it is just the beginning. All functionality of the solutions is built on top of it. Do not be fooled. Do not try to meet your requirements by extending standard CRM or ERP with bespoke development. For a robust and yet flexible, extendable and maintainable solution, you will need a solid information model.

Key Data Elements

In general two types of information are identified:

- Master data – this information describes the customers, connections, meters and associated elements
- Variable data – this includes information such as
  - Meter Readings
  - Consumptions
  - Tariffs
  - Invoices and Invoice lines.

Master Data

Mapping the physical reality to CIS (Customer Information Systems) or MDM (Meter Data Management) software is what the 'master data' part information model has to offer. All entities linked to delivering a commodity to households or enterprises are modeled in the EUCA model. These include connections, assets (such as meters) and channels which are used to store metering values.

To explain the flow of information through EUCA we will show you some examples.

Simple household

The physical situation in this example is a single household, with a basic electricity connection, a simple electricity meter and only one electricity tariff for all consumptions periods.

Figure 11 - A simple connection

There is one physical connection for delivering electricity to this residence which will be mapped to **one connection** object in EUCA.

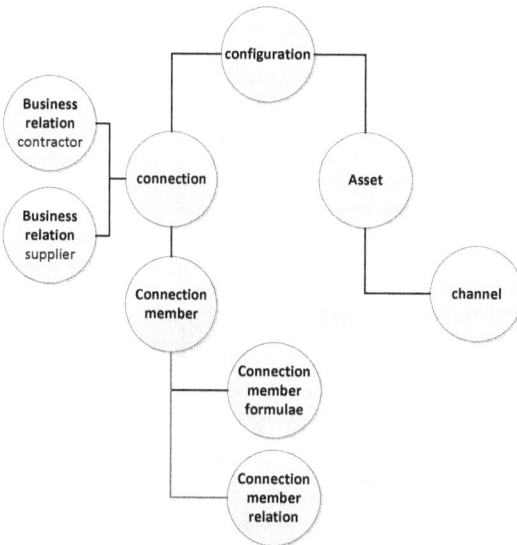

Figure 12 - Grid Connection Model

The connection has an address which in this case is the address of the house itself. It has two time-sliced **business relations** attached to it: the **contractor** which is the person in this household who is paying the bills and the **supplier** which is the market party who is billing the electricity to this household.

The actual consumption which is calculated based on the meter values, is stored in the **connection member** representing consumption total hour.

The electricity meter is a meter with one register measuring the consumption on a continuous basis. This will be mapped in EUCA by using **one asset** (which is the device) with **one channel** (in which the metering values are stored).

The asset or device also has an address which can be the same as the address of the connection and the house itself. The second attribute of the device is the channel which is in this example a single register which contains the meter values for the total hour consumption of electricity.

To link both entities together and use them in the application to represent the actual physical setup in this house, we will use the **configuration** entity. In this configuration, we will link the connection with the device and map the channels with the correct connection members. To calculate the consumptions, the connection will attach "**connection member relations**" to the connection members. These connection member relations will map to certain registers in the configuration and will be used together with the formulae (attached to the connection members) to calculate the actual consumption.

More complex situations

During the design of the EUCA model, flexibility has been top-of-mind. The rise of smart metering, complex metering situations in enterprises or split billing with sub-metering, EUCA can always model these situations in a highly performing data model.

More complex situations such as the use of different tariffs (day / night) for consumptions of the same commodity are handled through the use of multiple connections members in one connection. Each connection member then stores the consumption for this specific timeframe.

Configurations can contain multiple connections and multiple devices or assets, to allow sub-metering or handling multiple

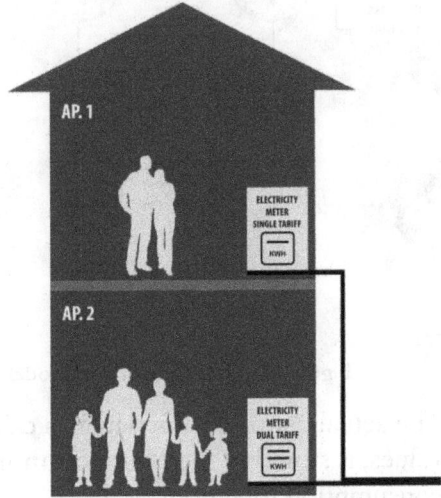

Figure 13 - Connection/Metering Configuration

connections for one owner. All building blocks of EUCA can be used, re-used and linked together to represent every possible scenario in the physical energy and utilities world.

This flexibility ensures utilities can incorporate future changes into their system with mere configuration instead of complex additional development.

Blueprinting

Although several combinations of connections and assets are possible, for a large utility only a limited number of configurations keep arising. For example: a household with a single tariff gas meter and a single smart meter for electricity.

The concept of 'blueprints' is introduced in EUCA to increase efficiency when introducing master data for new customers.

Blueprints are predefined models of reality which are stored in the system and are instantiated when a new customer is entered. Instead of creating all separate entities such as the meter device, the connection with the particular connection members, linking them using the same formulas etc., the MECOMS™ operator chooses the correct blueprint for the specific physical situation and MECOMS™ creates and links all entities together hence increasing efficiency dramatically.

Variable data

Metering data and load profiling

Despite the roll-out of smart metering, the vast majority of (residential) meters is still being read manually. This means the utility only receives metering values once every year. However, to allow efficient management of resources (forecasting of loads, purchase of energy etc.) a utility needs a more continuous view on the consumption of its customers. For these residential customers, the concept of SLP's or Synthetic Load Profiles is used.

Based on specific characteristics of the consumer and the actual measured consumption over a period, the system derives a consumption pattern of the consumer which is more realistic than simple linear extrapolation of the yearly measured consumption. SLP takes into account factors like for example: higher consumption of gas during winter periods or less consumption of electricity during the night.

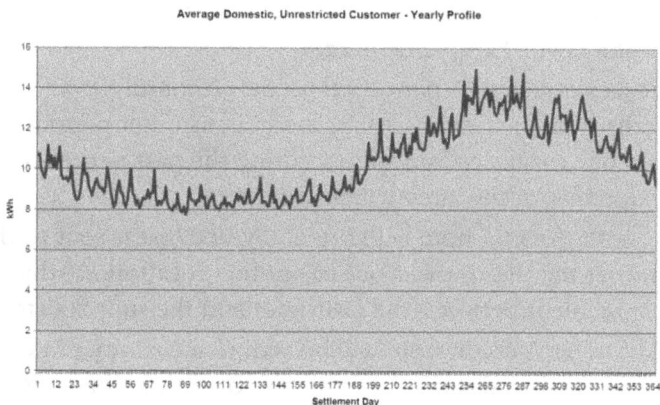

Figure 14 - Real Load Profiles

When grouping customers together, statistical analysis shows the deviation from actual load profile becomes minor.

For automatically read meters or AMR's, there is no need to use SLP's, since the 15 minute values present an actual load profile (RLP)which can be used to drive forecasts and calculate loads and peaks in the network.

Tariffs

For each consumption linked to a specific period, a tariff is required to calculate the cost (or profit) related to a consumption (or production). Once the consumption has been calculated on each connection, the tariffs will be used to determine the amount to be charged or credited.

These tariffs can change according to the amount of consumed energy. For example, the price will be less if the customer consumed less than x amount or it can be higher if he has consumed more than y. This is called volume slicing.

Invoice lines and invoices

Many different types of invoices are possible. The billing frequency defines how often a connection will be billed.

- Advance note: Advance notes allows the customer to spread the payment of his periodical invoice through advance notes. The amounts already paid in advance will be deducted from the total amount of the periodical invoice. An example of advance notes is monthly advance notes, with one closing invoice yearly.
- Manual invoice: A manual invoice is used to credit a one-time cost or a cost of a specific item.
- Periodical invoice: A periodical invoice (monthly, every 3 months, yearly) contains the amount a customer periodically has to pay for his consumption during the past period minus the amounts of the already paid advances.
- End note: An end note is the final invoice that is sent to the customer that has moved out to another location. At that time, the agreement between the customer and the supplier stops.
- Credit note: A credit note will be sent to a customer in case he has been invoiced the wrong amount or if something went wrong during the billing process. The credit note rectifies this error.

An invoice is created based on pre-calculated invoice lines which are the result of a billing calculation in monetary value (€, $...) for 1 item in the product, for example the electricity consumption in peak periods.

Important to point out is that the system stores every line in a calculated invoice lines table, as soon as information is available in the system (i.e. the meter readings are fed in from an AMR head-end). Only when the invoice is actually generated, the calculated invoice lines are collected.

This way, the actual billing run only needs to grab the invoice lines and consolidate them into the actual invoice, resulting in more error-free billing runs and more efficient use of computing resources.

Time slicing and versioning

This is an essential design element of the MECOMSTM database. Almost every entity in MECOMSTM is time sliced.

Time slicing is a way of versioning data. MECOMS™ does not delete any data. Data is time stamped which makes it possible to keep track of the history of objects and their related data. This way, time slicing allows the introduction of new entity versions without losing the old versions. To be more precise, if a new entity version is introduced in MECOMS™, the previous version is given a valid until date. The new entity gets a valid from date. Thus, the time slicing offers the possibility to reconstruct a situation in the past or it can indicate, for example, that a connection will be active from a certain start date (valid from) until a certain end date (valid until). In case no end date is specified, the entity is considered to be active indefinitely.

Entities which are time sliced are, amongst others:

- connections
- configurations
- devices
- tariffs
- formulas

Time slicing is also important for registering consumptions. In the utility market, it is important to register the consumption at different moments of the day. Each of these time sliced consumptions can be linked to certain tariffs for billing purposes. In addition, time slicing also allows for regressive calculation (performing the calculation of an entity as if the calculation happened in the past = flash-back functionality) or performing calculations as if they would happen in the future (forecasting purposes).

This concept is vital in deregulated markets in order to be able to reconstruct situations in the past.

☒	Question	Why this matters
☐	4.1.1 Are all the Utility related data elements available in one system?	Many solutions do not have an integrated data model. Data is stored in various systems. This results in serious headaches when it comes to integration and reporting. The MECOMS™ information model and therefore also the data model is stored in one database.
☐	4.1.2 Is the mapping of data elements with your business context adequate to implement the business requirements?	People tend to forget that there is a close relationship between business processes that need to be implemented and the availability of data. One can design a state of the art process but it will only work if the required information is available. It is therefore essential to ensure that the information/data model can support your organization.
☐	4.1.3 Can the Solution work with high volume time series typical in Meter Data Management?	The context of smart (-er) solutions has already been described in the introduction chapters. Regardless of the legislation requirements, storing meter reads is highly storage intensive. This thought process will give you a table like this:

Meter Data Collection

Customer Type	No of Customer	No of Meters	No of Channels	Time Interval	Collection Method
Industrial & Commercial	N1				
	Nx				
Residential	N3				
	N4				
	Ny				

For example: N1 x 3 meters 3 channels x 96 intervals (a meter reading every 15 minutes) = 864 meter readings per day, for only 1 customer. What if you have millions of customers?

☒	Question	Why this matters
☐	4.1.4 Do we have a clear view on meter data storage requirements?	Consider: • The number of meters you need to support. • The time interval for data to be retrieved from the meters. • Any real-time data retrieval requirements. Some legislation rely on "day after" data collections. The trend is to get 15, 30 minute or hourly values in real-time. • What is the differentiation of customers? Are there differences in time intervals for data collection for Industrial & Commercial and residential connections? • Which information do you want to collect from the meters? Consider electricity, gas, water and/or thermal integration with one gateway. The number of channels you want to collect could include time of use channels (day, night tariffs), active, passive, reactive energy, peak energy and so one. • Which is the method of collection? Can all this be stored in the solution database? • Is it an integrated solution or is data stored in separate systems?

☒	Question	Why this matters
☐	4.1.5 Can the solution work with Time slicing?	Knowing the current situation of the grid is essential. The requirements will be different depending on your role in the market. But what if you need to know how the objects you manage were related in the past? This includes meters, customer, connections etc. Examples: • What is the serial number of the meter installed at a connection one year ago? What was the meter read? • Which customers (net users) historically paid the consumption on a particular net connection? • Which tariff was used to create an invoice 2 years ago? You need a solution that supports time slicing to answer these questions. MECOMS™ supports time-slicing in a highly intuitive and user-friendly way.

☒	Question	Why this matters
☐	4.1.6 What if data elements are missing in the solution?	The advantage of a standard solution is that it contains years and years of experience. This way you can maximize the out-of-the-box functionalities offering you not only a cost effective but also an upgradable solution. The goal should be to use the standard out-of-the-box solution as much as possible. But what if bits of information are not represented in the information model. Consider two situations: • To meet your information requirements you need an extra attribute in a data object. Example – You want to store the color of the meter. This should be pretty straightforward to add. • A data object is missing. This requires a data table to be added to the database. This of course is fully supported by the MECOMS™ ™ architecture. The MECOMS™ property framework allows you to add attributes without programming. Tip – think twice on how to fit this into the standard information model.

Chapter

5

Utility Processes Model

Everything should be made as simple as possible but not one bit simpler.

Albert Einstein (Physicist, 1879 - 1955)

MECOMS™ and business processes

HOW can we use the functionality offered by MECOMS™ in real life situations? Thinking about functionality alone is not sufficient.
We need to look at the business processes required to meet the business objectives. Business processes are collections of related, structured activities or tasks that produce a service or product to execute the company's mission, vision and strategic business objectives.

Figure 15 - Developing the right processes

Understanding the business model

Using Business Process Modeling techniques (swim lanes, BPM, etc.) business leaders, analysts and managers can analyze and improve processes in the organization. Having standard business processes available is a huge help when optimizing the time and therefore the cost of implementation. But before you start worrying about business processes you should first think of the company's business model.

Business processes are defined to meet the requirements of the company's business model. We recommend using the Business Model Canvas method to get an excellent insight in the business model. It will show you how the company wants to operate. This gives an excellent insight in how to tailor the business process to execute the vision of the company. The resulting Business Model Canvas gives you an overview of the market proposition of the organization by detailing:

1. the **customer segments** the company wants to work in,
2. the **channels** toward the customer segments we want to develop,
3. how the **customer relationship** is developed and maintained.
4. how all this will generate **revenue** for the utility

Central in the Canvas is the **value proposition** the utility wants to bring to the market. The canvas also looks at how this will be done by defining:

1. the **key partners** required to support the business.
2. the **key activities, processes** the utility needs to execute (This gives us a fair insight in the company's main processes)
3. the **key resource** to make the business model work
4. the related **cost structure**

The Business Canvas Modeling is a technique that can be successfully used in brainstorm sessions to define, refine and explain the companies approach to business. It allows utilities to create business models that differentiate them from the mainstream. This is the input needed to build the business process insight required to set up a solution.

The following figure gives an example of a hypothetical utility company in the utility market, called Omegapoint. It demonstrates how you can gain a clear insight in the companies operation on one page.

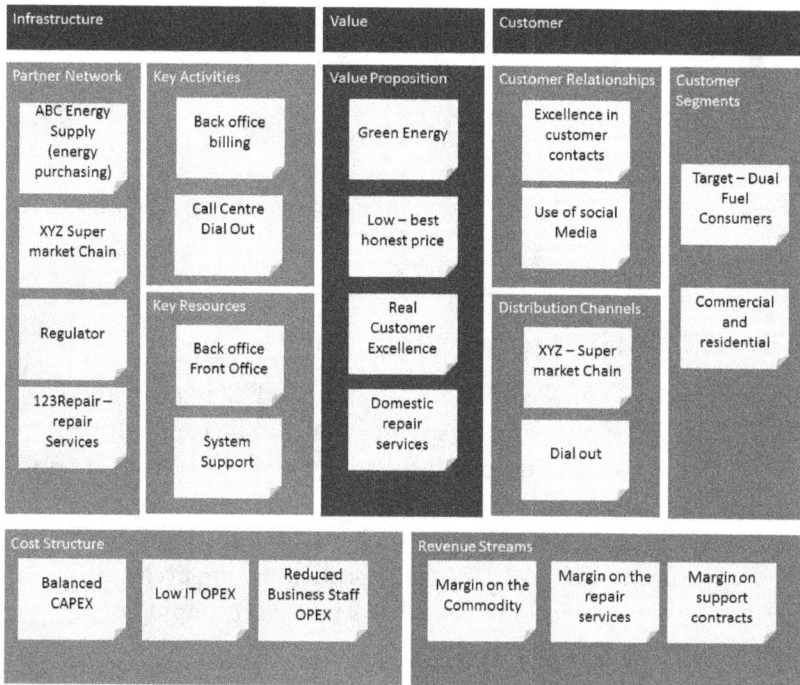

Figure 16 - Business Model Canvas - Understanding the Business of fictitious Omegapoint retailer in a deregulated market

The standard MECOMS™ Model and business blueprints

Once the business model is clear, developing business processes becomes a much easier job.

MECOMS™ comes with a standard process catalog. This is what we call the MECOMS™ Process Wheel, MPW. The MPW covers typical Utility processes and offers a blueprint process.

Figure 17 - The MECOMS™ Process Wheel as process library, roles, application role center

The MECOMS™ Process Wheel, MPW

The MECOMS™ Process Wheel, MPW is a catalog of standard blue print processes that can be used out-of-the-box or can be modified to differentiate the way of working. In a heterogeneous environment, processes often need definitions across systems which obviously complicates matters. The MPW contains individual life cycles of business objects which allows for easy grouping of processes in the catalog.

Processes

Standard MECOMS™ processes are defined as swim lane process diagrams. These are standard blueprints processes that can be implemented, changed, enhanced, enriched and simplified to meet the business requirements.

Roles

People in the organization are linked to roles to represent various job functions. These roles tie into the security system. We have identified some 20 standard Roles to be used. They can be changed and new roles can be added depending on the organization's needs.

Application Role Center

With the MECOMS™ application each role has its own role center with work queues, KPI's and relevant reports. The role center is the prime page for users in the MECOMS™ solution.

Fit/Gap

Many of the business processes used in utility companies are the same or at least similar. In a deregulated environment a reduction of the "Cost-to-Serve a utility customer" is a competitive advantage. This is often related to business process optimization and differentiation from the competition.

There are bound to be gaps between the standard MECOMS™ product and a company's business model and requirements. The MECOMS™ product and the MECOMS™ Sure Step Implementation methodology come with tools to ensure that the fits and the gaps can be clearly identified.

The MECOMS™ Process Model

MECOMS™ Process Wheel

The MECOMS™ Process Wheel consists of a number of different elements that make up the process model. This allows grouping the individual processes in a logical way.

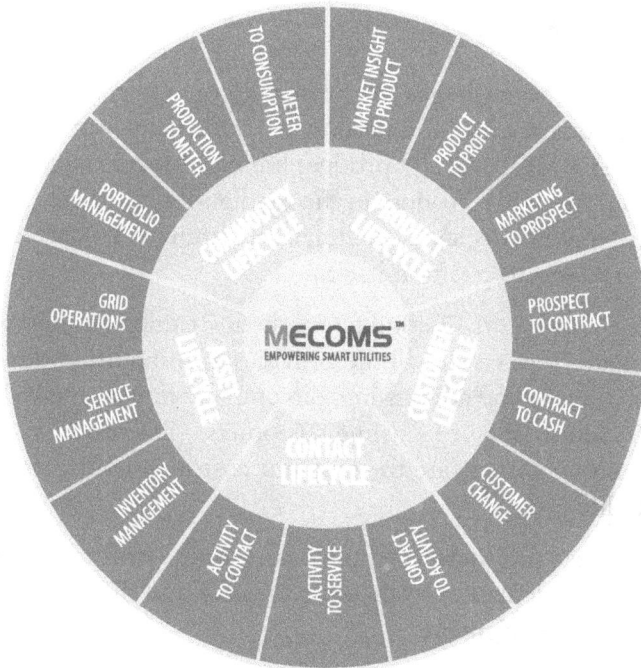

Figure 18 - MECOMS Process Wheel, the Business Process Library

Commodity Life Cycle

How are we dealing with the delivery of the commodity to the customer or what we call the Commodity Life Cycle?

MECOMS™ ™ deals with non-operational processes of the utility. A utility needs to have processes for forecasting and managing the demand. It must organize a purchasing and trading element for this.

Obviously it is far more difficult to deal with this when a distributed production is in play. This theme mainly relates to electricity. It requires differentiation between production and consumption. Sometimes this is called Presumption.

Within the commodity life cycle a key aspect is Metering. Metering is an art, a real meticulous craft dealing with data collection in various manual and automatic ways. Once the raw meter reading data is collected, a number of calculations are required after which we need VEE: Validation, Estimation and Editing. Validation of the correctness and plausibility of the reading, Editing for changes and Estimation in case no reading is available.

Product Life Cycle

A set of processes manages the product(s) the utility is dealing with. This could be any commodity, gas, electricity, water, thermal energy… The MECOMS™ product life cycle covers all processes related to putting the product in the market. Based on market analysis, the utility decides which market segments to address and designs products to position itself.

Then the utility wants to conduct marketing campaigns and create market share. This covers all the activities to bring the product to the market, including advertising, direct marketing through the use of CRM and upselling of existing customers. The products being sold to your customers are the basis for customer contracts in the customer life cycle.

For all of these products, the utility assesses the profitability, including the calculation of the cost-to-serve and the associated revenue for each segment.

Customer Life Cycle

The Customer Life Cycle covers all aspects of the relation with a customer, from acquisition to deregistration.

The first step in the Customer Life is when prospects have signed up and become actual customers. This includes creating agreements, handling the move-in and any future changes to the contract with the customer.

With the customer's contract and consumptions as input, the contract to cash processes generate the actual revenue from the customer. This includes billing (paper-based or online), collections and credit management.

Customer change is crucial to any service delivery, so the Customer Change process takes care of any changes to the customer's data, including customer moves and switches.

Contact Life Cycle

The Contact Life Cycle groups all processes related to interaction with the customer.

In the Contact to Activity process, customer enquiries result in activities which are used to efficiently respond to the customer. These enquiries can be phone calls, emails or web based communications.

Some contact activities need technical interventions and will result in a service request. This will trigger a process in the asset or commodity life cycle.

Processes from other life cycles can also require a utility employee to initiate a customer contact. The activity to contact process aggregates these requests and provides generic processes to handle these interactions.

Asset Life Cycle

The asset life cycle combines all processes related to the managing of physical assets. In a utility context we manage metering devices, grid components etc.

Inventory Management contains all processes regarding inventory, human resources, vehicles and car warehousing from procurement to stock.

The Service Management process groups all activities to bring stock assets to site for activation, to maintain assets during their life span and to decommission assets to stock or scrap.

Grid operations cover the day-to-day use of assets. In a smart-grid setting, these processes cover load-limiting, activation/deactivation, outage management and communication through the asset tree.

☒ Question	Why this matters
☐ 5.1.1 How is the fit gap? How efficiently can the processes be executed?	Consider making an inventory of all the individual processes and all the end-to-end processes you want to run. Which are the functions that the processes are going to use? Are they flagged available (Fit) in the functional fit/gap list? People execute processes so it is good practice to check whether the end-to-end processes can be executed efficiently. This gives you a good impression of the work that needs to be done in configuring and implementing the processes.
☐ 5.1.2 Can the processes cope with the volumes of work?	Processes need to be designed to handle the volumes of work required. Consider the people workload of the processes you design. Does the human workload make sense? What is the cost perspective? Consider the load on the IT architecture of the processes you design. For batch processes that run in the background, you need to develop a schedule when they are planned to run.
☐ 5.1.3 Are we not automating too much?	Example – automatic correction processes do not make any sense when a utility only serves a few (large B2B) customers. On the other hand if the utility is dealing with millions of customers, a manual process will prove to be unfeasible and too costly in terms of labor cost.

☒	Question	Why this matters
☐	5.1.4 Are the Business Processes designed efficiently?	Often a variety of processes are in place. If you really want to lower the cost-to-serve, you need to carefully look at how you can optimize the process at a minimal cost and maximal customer satisfaction. Think about basic total quality management (Deming, Crosby, lean) concepts such as the cost of quality (the cost of doing things right the first time – the price of conformance) and the cost of poor quality (rework, dissatisfied customers, loss of income – the price of non-conformance) Keep processes lean!
☐	5.1.5 Are processes well documented?	Many organizations do not have documented processes. They rely on domain specialists who are typically overloaded and handle Price of Non-conformance/Cost of Non-Quality items. You can use, extend and change the MECOMS™ process wheel and process swim lane diagrams and associated documentation.
☐	5.1.6 Are you considering e-learning material for dealing with knowledge management?	By creating e-learning material and screen recordings, you make onboarding of new staff much easier.
☐	5.1.7 Are you considering end – to – end processes?	System users, consumers, business partners experience end-to-end processes. So, think end-to-end processing.

☒	Question	Why this matters
☐	5.1.8 Are you considering the relation between the information model and the functional model?	Processes rely on functionality and data to be available. There is a clear relation between the information, functional and process model. When designing processes keep this relation in mind. Also consider where data and functions reside. This might be in an external system that can be addressed using a Service Oriented Architecture.
☐	5.1.9 Do you have a strategy for exception handling?	Processes do go wrong. There will be exceptions that lead to incorrect or inconsistent data. So you need to consider how to correct these exceptions. MECOMS™ has well defined mechanisms for exception handling.
☐	5.1.10 Are you planning for process testing in your testing scenarios?	It is good practice to include process tests in the test scenarios. In order to carry out realistic end-to-end process tests make sure representative test data is available in your test system and environment.

☒	Question	Why this matters
☐	5.1.11 Did you develop your business KPI's based on the process performance?	Your company has a mission, a vision and strategy. A balanced scorecard can be used to track strategic objectives related to strategic themes in a strategic map (Check for information on Kaplan's balanced scorecards and strategy maps) Each strategic objective has one or more corresponding measurements. MECOMS™ allows using the role center to visualize traffic light KPI's with trend analysis on process measurement that you have defined. It is good practice to design KPI's at the same time as the process. Do not postpone this. You cannot run processes without knowing the required performance. Understand the process interdependencies so as to determine the impact on the strategic results you have set.

Chapter

6

Systems Architecture

Architecture is the art of how to waste space.

Philip Johnson (US Architect, 1906 – 2005)

Systems Architecture and MECOMSTM, an Overview

W HEN building an IT solution, the system architecture is a crucial aspect. Systems architecture is not a new discipline it has been around for thousands of years. If you want to build a house or a civil construction, you start with understanding the requirements and you carefully plan floors, rooms and other structures to best meet the requirements and ensure the structure is easy to maintain and expand.

The same applies to IT systems architectures. This makes system architecture a wide–ranging discipline. You can look at systems

Figure 19 - System Architecture

architecture from various angles:

From a Business Context:

- Data
- Business process and Logic
- Performance and Sizing

From an IT Context

- IT components and functions
- Virtualization and Cloud
- Security
- Performance and Sizing (capacity management)

Holistic Context

We strongly believe you should not look at system architecture from an IT perspective only. IT is there to support the business. Forget all the IT Fuzz; think of what the business requires. What are the system architecture components of smart utilities solutions?

IT Systems do not exist in a vacuum. They serve the vision, strategy and operating model of the organization. But often aligning the goals of the IT Systems with the business requirements (as well as the organizational alignment) can be a challenge. Operating models change under pressure of the competition, market, legislation... The challenge of the IT solutions is to swiftly follow the changes as they are essential to execute the vision.

This requires a solid architecture of IT systems. Enterprise architectures are reflecting the integration and standardization requirements of the company's operating model. It is the basis for building business processes and IT infrastructures.

MECOMS™ on Microsoft Dynamics AX offers a multi-tier architecture that provides the foundation for execution and growth. It is a solution designed to close the gap between the ever-changing business requirements and the IT system. The key focus for effective enterprise architecture is on data, transactions, processes and customer (both internal and external) interfaces. They ensure that the value stream vision of the company's operating model becomes reality.

The Microsoft SERA Architecture

Microsoft Utilities has developed a reference architecture for the Energy and Utilities industry. It is called SERA, Smart Energy Reference Architecture. The SERA document describes a proposed infrastructure and application landscape for utility companies and elaborates on how to manage, monitor, control and report the assets of the new smart energy ecosystem. This includes both IT assets and utility grid assets.

The SERA Reference architecture looks at:

- **Evolution of the Grid**. This section describes the forces shaping industry direction and intends to provide an overview of the challenges coming to the fore.
- **Changing Demands on the Business**. This part offers an industry architectural vision elaborating on the entire value chain from the utility to the end-use consumer, whether they are commercial, industrial or residential. Business decision makers will gain a greater understanding of the business challenges they will face as the smart energy ecosystem emerges.
- **Architecture.** This section will be most useful to software developers, system integrators and solution specialists who already have an in-depth understanding of the industry and information architecture and are mainly focused on Microsoft technologies.
- **The Microsoft Technology Stack**. This section identifies Microsoft products and solutions, as well as partner-led solutions in some cases which enable this architectural vision.

The MECOMS™ business solution fully aligns with the Microsoft SERA architecture. In this chapter, we will look at some top level aspects related to the MECOMS™ solution.

Maturity Model and Microsoft Optimization models.

Microsoft has developed a systematic process for assessing the maturity and capabilities of system architectures, including application platforms to build a dynamic IT architecture. The models help to align the IT Solution with the business, making IT a strategic business asset to drive innovation. The optimization initiatives put the user, the people driving the business in the center.

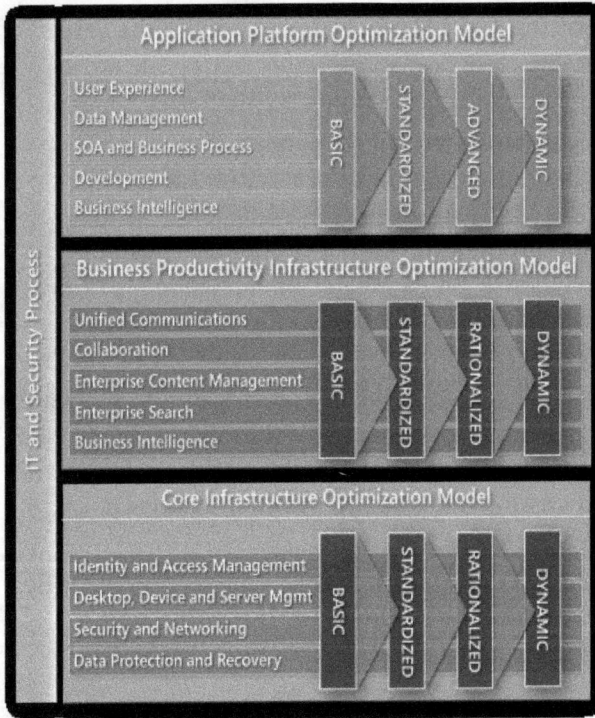

Figure 20 - Microsoft IT Maturity Model

Microsoft calls this "People Ready Business". It is about leveraging the biggest assets an organization has – the people. The people in an organization are the core and driving force in achieving the business outcome. MECOMS™ embraces this vision with its focus on the usability of the solutions and extending the core Microsoft Dynamics AX architecture that is built on the dynamics IT principles.

The Microsoft optimization models have been developed using industry best practices. They use a four level maturity model that can be used for assessing the maturity of the core infrastructures, business productivity infrastructure and application platform.

Level	Characteristic	Details
Basic	Cost Center	Uncoordinated, Manual infrastructure, Knowledge not captured
Standardized	More Efficient Cost Center	Managed Information Technology infrastructure with limited automation and knowledge capture
Rationalized	Business Enabler	Managed and consolidated information technology infrastructure with extensive automation, knowledge captured and re-used
Dynamic	Strategic Asset	Fully automated management, dynamic resource usage, business lined SLA's, knowledge capture automates and use automated.

MECOMS™ builds on the huge 9 Billion$ yearly Microsoft investment in the platform and architecture. This allows the Ferranti Product Development organization to focus on what Ferranti is best at: building great Utility Business Applications.

Microsoft Dynamics AX and MECOMS™ Core architecture

Overview

The great thing about the MECOMS™ architecture is that it extends the Microsoft Dynamics AX architecture. This allows the MECOMS™ product development teams to focus on functionality rather than the technical infrastructure.

Microsoft Dynamics AX has been designed for the enterprise. It comes out-of-the-box as an integrated infrastructure. You do not need to worry about integration, it just works.

The logical architecture describes the various components that define a Microsoft Dynamics AX environment. This does not mean that the physical infrastructure needs to be completely the same as the logical architecture. It is up to the technology consultant and architect to scale the architecture and determine the number of physical (or virtual) servers required for each of the components of the architecture.

Microsoft Dynamics AX and therefore also MECOMS™ uses integrated Windows Authentication to authenticate Windows Server Active Directory Domain Services users. You can configure Microsoft Dynamics AX to use other authentication providers (example LDAP)

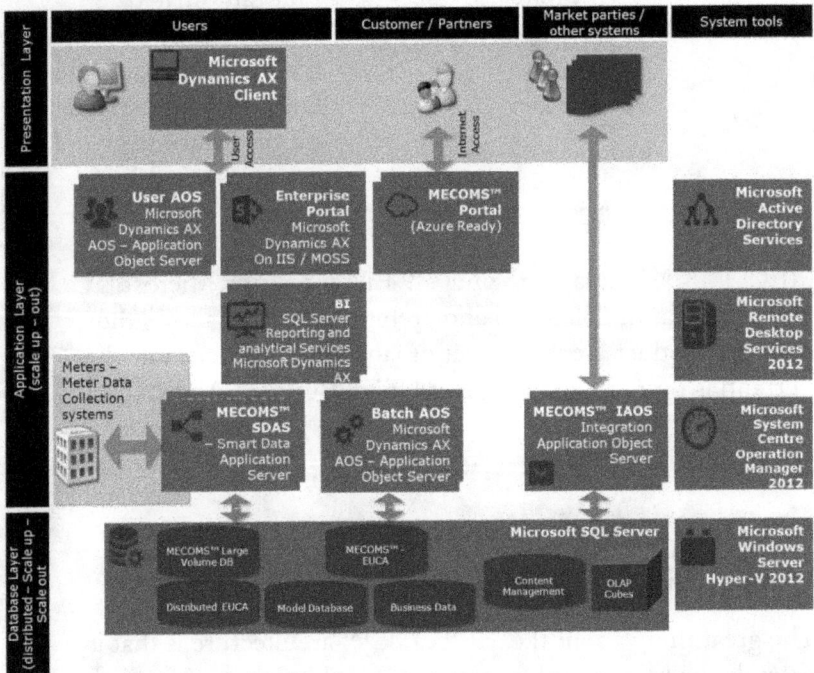

Figure 21 – MECOMS™ - Microsoft Dynamics AX Architecture

Within Microsoft Dynamics AX the security is governed to authorize access to data, functionality and presentation elements such as forms and reports.

As MECOMS™ extends the Microsoft Dynamics AX architecture it can make full use of all the goodies like:

- The standard ERP data structures and functionality
- Security and access control systems
- The Microsoft Dynamics AX application object server
- The reporting services infrastructure
- Business Intelligence infrastructure
- Thick and thin clients with role centers.
- Intranet and internet group service using Microsoft SharePoint.
- The web services framework offered by Application Integration Framework
- Office application integration.

System Components

Presentation tier (clients and external applications)

Users in MECOMS™ can be internal business users or external customer and business users. As MECOMS™ extends the Microsoft Dynamics AX platform it uses the same presentation layer functionality as AX. In the presentation tier, a client provides a user interface to Microsoft Dynamics AX and MECOMS™ functionality and data. Clients can also be external applications incorporated in the architecture to integrate functionality or to exchange data. These external applications can reside within the company or can be Business-to-Business connections with systems of business partners.

In general following interfaces are available:

- The thick windows client for Dynamics AX is a native windows program that provides the user with a rich user interface.
- The Enterprise Portal provides access to functionality and data via web browser.

- To build interfaces with external applications you can use the Application Integration Framework. The Application Integration Framework provides an extensible framework for XML-based scenarios for Enterprise Application Integration (EAI) and a Service Oriented Architecture (SOA)

Application tier

The application tier consists of Microsoft Dynamics AX components that can be scaled out to meet performance requirements as well as functional requirements.

A number of different components are identified:

- The Active Directory Domain Controller – This component controls user authentication.
- The Users and Batch Application Object Server (AOS) – This component executes business logic of the business applications. It controls the communication between Microsoft Dynamics clients and the database. AOS's can be configured as single instance or load balance clusters of multiple instances running on multiple servers. This allows for parallelism in business logic execution. The AOS also enforces the application security. The AOS communicates with clients using AOS services and also runs the workflow. This allows you to create individual workflows or business processes. Batch AOSs run background processes such as billing.
- (optional) Enterprise portal – The platform provides a set of websites that give you access to data and interact with business process using web forms. The enterprise portal uses the content and configuration management of Microsoft SharePoint or SharePoint services.
- The (optional) Business Intelligence servers run:
  - Reporting – Microsoft SQL reporting is the primary reporting platform for Microsoft Dynamics AX and MECOMS™ and comes integrated out-of-the-box. This does not stop you from using other reporting tools.
  - Analytics – The Microsoft SQL Server analysis service is a server based environment integrated in

Microsoft Dynamics AX that allows for online analytical processing (OLAP). The analytics trend reports and KPI's give business insight to the user. The Microsoft Dynamics AX Analysis Services project wizard makes it easier to create, update, configure and deploy Analysis Services projects. The wizard performs many of the tasks that you would otherwise have to perform manually in Microsoft Business Intelligence Development Studio (BIDS). The tooling comes integrated out-of-the-box. Similar to reporting services, this does not stop you from using other reporting tools.

- The (optional) MECOMS™ integration Application Object server runs the integration framework – This component offers the capability to integrate Microsoft Dynamics AX/MECOMS™ with other systems inside or outside the enterprise. Data Exchange is leveraged through formatted XML files.

- The (optional) MECOMS™ portal is a .NET deployable self-service portal for utility customer that can be deployed on premise and in the Microsoft Azure cloud.

- The (optional) MECOMS™ SDAS, Smart Data Application Server allows for a high volume, high bandwidth connection with Smart Metering and AMR infrastructures. The data can be stored in the Microsoft SQL Server MECOMS™ relational EUCA model or in the Microsoft HDinsight Big Data Large volume database.

Other Optional components include:

- Microsoft Remote Desktop Service 2012 to deploy thin clients.
- Microsoft System Centre Operation Manager 2012 tools for system management.
- Microsoft Windows Server Hyper-V 2012 provides an environment for efficient low cost virtualization of servers.

Data tier

Microsoft Dynamics AX and MECOMS™ use the Microsoft SQL Server database. The relational database store is used. MECOMS™ can use the big data components of SQL Server for dealing with high volume meter data management requirements. Database replication and clustering techniques for providing data redundancy and scaling can be used.

To deal with large volumes of data, MECOMS™ uses the Microsoft HDinsight Hadoop implementation for Large Volume Databases. The MECOMS™ EUCA database can be distributed to allow scale out in distributed database environments.

Management Tools

The Microsoft System Center Monitoring Pack for Microsoft Dynamics AX 2012 provides an end-to-end monitoring solution for Dynamics AX 2012. The management pack automatically discovers the entire AX environment including the databases, reporting servers, analysis servers, enterprise portal server and application frameworks and monitors each of the components for configuration, availability and performance. The monitoring pack provides early warnings that an operator can use to proactively identify issues that could affect the performance and availability of the Microsoft Dynamics AX system.

Object Orientation

The model store is a separate part of the Microsoft AX database where all application elements for the Microsoft Dynamics and industry verticals such as MECOMS™ are stored.

All layers below the ISV layer contain the standard ERP functionality which can be extended in the upper layers to offer specific and vertical-focused functionality.

The MECOMS™ out-of-the-box application is built in the ISV layer which extends several functionalities of the lower layers, like customers, invoices etc. Because of the API's between the different layers, a change in the lower layers will not have an impact on the higher layers, as long as the API remains the same or remains backwards compatible.

Because in certain project or geographies, MECOMS™ customizations will be created by people in our partner channel, we can segregate the responsibilities by allowing these customizations in the VAR and CUS layer. All modification to these layers will extend functionality out of the MECOMS™ layer.

Specific user configuration will be covered in the USR layer, where modifications on an individual basis are stored.

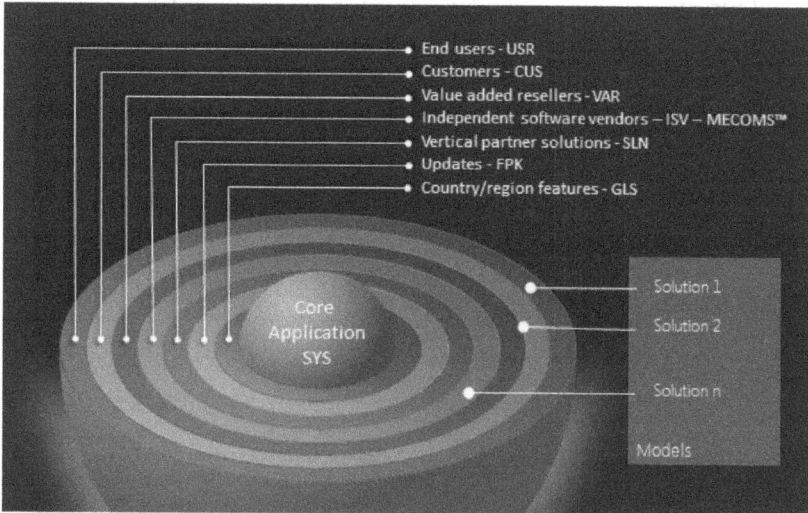

Figure 22 - Application Model

Applications in the Cloud

Microsoft strategy is to move applications to the cloud. Some components such as MECOMS™ portal can already be operated in the Azure cloud.

User Interface

Utility Solutions have various users. If you want to live up to lowering the "Cost-to-Serve" objective you need to make sure you build a user-friendly solution. Users of a utility not only include the internal business user. You also need to consider the external users like customers, consumers and business partners. By offering a solution that meets their business needs that helps them to do their tasks and give an insight in KPI's and trends related to their processes, we can dramatically increase productivity.

Utility customers and partners

Both industrial and residential consumers will want to have access to utility data using a web portal. They might want to use social media. They might want to use a voice response system to interact with the utility.

These are examples of customer interaction channels with the consumers and partners. In principle they can use any media. One could interface using the television setup boxes or kiosk. Regardless of the platform, the underlying technology uses robust web services that interface with the Microsoft dynamics AX Application Integration Frame work. Typical functions are bill and consumption profile Presentment.

An example of such a portal is shown in the next screenshot from the Omegapoint demo.

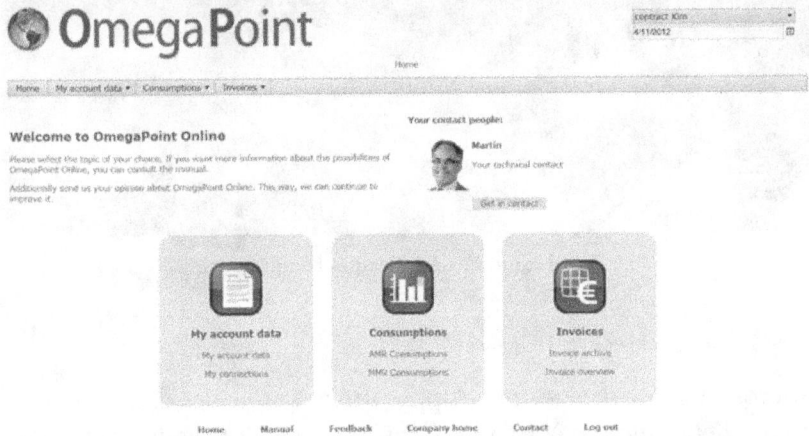

Figure 23 - MECOMS™ Webportal

Utility Users – Role Center

The enterprise portal infrastructure comes with role-specific home pages that are called role centers. The role center is the home base for a particular role in the organization. The role centers can be freely configured. Obviously it is the objective to define roles and associated role centers that meet your organizations requirement. The daily information and transactional requirements of the CEO tend to be quite different from the information a call center employee requires. The Role center is in fact configurable with web parts that come out-of-the-box and that interface with the business application.

Web parts that can be added on the role centers include:

- Menu Structures
- 'My-Favorites' menu structures

- Quick links to web address
- Info from internet, intranet and extranet sites.
- Transactional data.
- Alert Links – that show you records that have configured alert conditions.
- Cues – These are like inboxes of work that needs be done. An example could be outstanding customer complaints.
- Reports – generated by the reporting subsystem and giving insight in the business performance
- Business KPI's and trends generated from the analysis services OLAP cubes, to give you a top-level insight at a glance

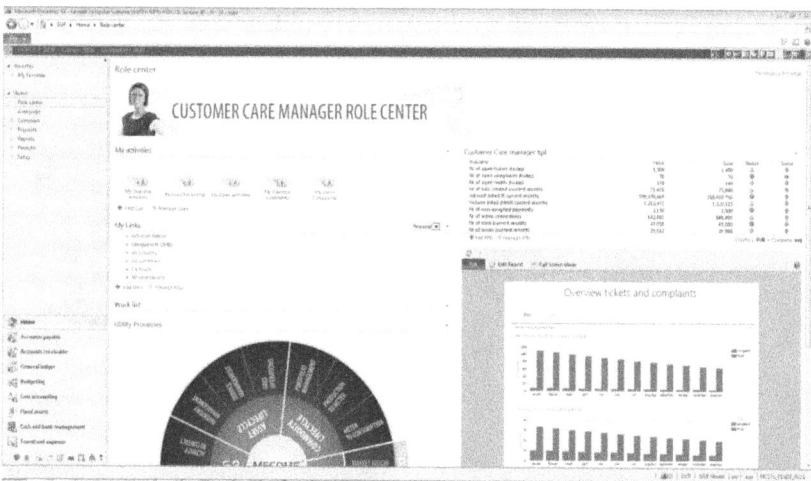

Figure 24 - Example MECOMS™ Role Center

MECOMS™ identifies a number of standard roles that are typically found in utilities organizations. Roles are fully configurable to align them with the organization that implements the solution.

Figure 25 - Standard MECOMS™ roles

MAREK
ASSET MANAGER

Challenges
- Keep track of all the assets
- Manage the asset tree
- Provide an easy integration between inventory and operational assets
- Manage purchase and service costs on assets
- Easily create new assets based on configuration

MECOMS™ Value
- Integration with Inventory Management
- Link to fixed assets and customers
- Smooth integration of working assets to service
- 360° view on asset related information

Figure 26 - Example MECOMS™ role

Utility Users – Transactional interface

A Microsoft Dynamics AX/MECOMS™ form is a window that shows information and allows you to perform actions. A common design pattern is used for all forms. A details form for example has following elements to ensure usability:

- File – Tabs Menu – allowing you to navigate the system and also to perform standard tasks such as Microsoft Office integration, record, alerting
- Ribbon Bar (known from applications such as Microsoft Office) – Action pane with actions to perform on the current record.
- Page with Fast tabs and with Action pane strip – shows you the various attributes/columns of a record in a group using the Fast tabs. Each Fast tab can have one or more buttons associated with it to allow for fast actions.
- FactBox pane – shows additional information related to the record that appears in the form.
- Status Bar – gives you record navigation and status information
- Help

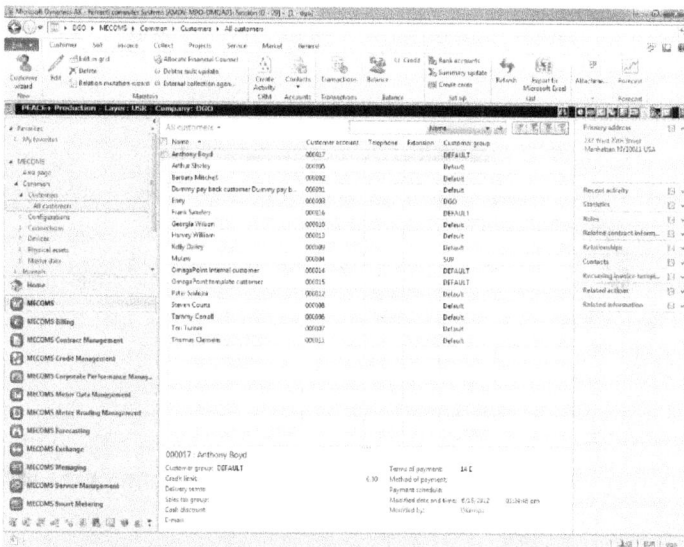

Figure 27 - MECOMS™ Transactional User interface

Sizing an ICT infrastructure, performance & capacity considerations

Introduction

Sizing an IT environment is never easy. There are a lot of factors to take into consideration. If you oversize the IT infrastructure, you are wasting valuable budgets. If you underestimate the ICT infrastructure, you will not meet the performance and capacity requirements of the project. It is very likely that this will result in loss of confidence in the project.

Sizing ICT infrastructures is about determining processing power, memory and disk capacity to meet the performance and capacity business needs. System architectures need to be developed to support all the Microsoft Dynamics AX/MECOMS™ architectural components you want to support.

The numbers of users that use the system obviously determines the system performance. Do not forget your self-service customer and partners. You should look at "an ordinary day in life" and determine how your infrastructure will deal with peak usage. The load will not only be generated by users. You also need to cater for batch processing (examples are Billing runs, meter readings). These batch processes also impact the load on the database, on application object servers and on webservers. It is always a good idea to develop a scheme for moving load to more quiet periods of the day.

To monitor all this, you will also need to consider how you want to manage the environment to give you information on resource usage and availability of the individual components.

Sounds Complex? It is! So use the reference architectures and benchmark data. They will help you in building the infrastructure that is right for your application.

Modern ICT environments are virtualized and operate in a private (or even public) cloud environment. The MECOMS™ solutions benefits from the entire Microsoft data center architecture and solutions to offer cost effective infrastructures.

A set of sizing tools has been developed to allow you to create an ICT environment that meets your needs.

Scalability

You also need to take into account the fact that the requirements might change. Think of the storage impact if one decides to store all meter reads, consumptions and invoices for multiple years to meet legislation requirements. This is the easy part, it is a business decision. But what if you operate in a deregulated environment and the number of customers your system needs to support might fluctuate (and even grow dramatically)?

Environments

The goal of the investment is to support the users of the application. As a result, you need a production environment that offers all the fancy functionality to the users.

You should consider having various environments to support the application maintenance/management process. Consider

- Production environment
- Acceptance environment
- Test environment
- Development environments
- Training environments

You also need to determine the dataset that is used for each environment. Is it acceptable to have an environment with a limited data set? Can you execute performance testing on a dataset with limited data records?

As you use the environments, the data is being polluted. Setting up a refreshing mechanism is required to know what the state is of the data you work with.

You also need to strategize on the time aspect. As time moves on, the meter readings, invoices, customer relations get old. This could impact your testing strategy. You might need to plan to insert new data.

High Availability and Disaster Recovery

The Application is the heart of the business operations. What do you plan to do in case of failure or disaster?

You need to consider building a disaster recovery plan. This will help you to know what to do when things go desperately wrong.

Modern infrastructures have fail-safe built in. This is also the case with Microsoft Dynamics. The underlying virtualization takes care of this.

☒ Question	Why this matters
☐ 6.1.1 Have you defined your Enterprise architecture?	It makes sense to determine how you want to deal with Enterprise architecture. Often this is a dictating document ("you shall not...") You need to think about architecture in a holistic way and you should also have processes in place to build the architecture. Look at frameworks such as Microsoft Operations Framework and Cobit and ITIL
☐ 6.1.2 Have you thought about the maturity level of your IT infrastructure and your business applications?	Using Maturity models and other tools to assess the maturity can help to develop your strategy. Examples of methodologies to do this are the Microsoft Core Infrastructure Optimization model, the Business Process Infrastructure Optimization model and the Application platform optimization model.
☐ 6.1.3 Does the System use a multiple tier approach in its architecture?	Separating user interface, business logic and data dramatically improves the manageability of the system. It also ensures uniform design patterns for individual instances of objects such as business logic and user interfaces. This also makes it easier to scale out the system to handle larger volumes and users.

☒	Question	Why this matters
☐	6.1.4 How componentized is your core infrastructure?	By using a component with clear functions you can build an infrastructure that meets your needs. This is far more flexible than a monolithic system that comes as a black box. By using the componentized flexible infrastructure of Microsoft Dynamics AX/MECOMS™, you design an architecture that meets your requirements. The advantage is you do not need to do the integration. This work has already been done by Microsoft. You choose, design, decide and implement the architecture with minimal effort.
☐	6.1.5 How does your application architecture deal with scalability?	Architectures should have the ability to handle a growing volume of work when faced with a growth situation. In ICT terms a system is scalable if performance and capacity increase by adding hardware. Scalability can be achieved by Scale-out (Scale horizontally) which involves adding more nodes, computers/servers. Alternatively an application can scale-up (scale vertically) which requires adding processing power, memory, storage to a computer/server. Microsoft Dynamics AX /MECOMS™ support both mechanisms.
☐	6.1.6 How does the system scale in terms of processing power for user support and batch processes?	MECOMS™ Scales by adding processing power by scale-out or scale-up both for user transactional processing and batch job processing.

☒	Question	Why this matters
☐	6.1.7 Does the system allow for audit trail of user action or system actions?	You want to know what has happened with the elements you are managing. An audit trail gives you a view on what has happened with each element and gives insight on who has taken a particular action. Audit trailing is an important function in the MECOMS™ application.
☐	6.1.8 How are you organizing the Connectivity to the outside world	Utilities applications connect with the outside world through web services, XML, EDI or CSV. You need to consider the security of these connections as well as the audit trail on messages sent to the outside world. The financial impact of market interaction should not be under estimated.
☐	6.1.9 How is the backup of data organized?	Most Utilities systems use advanced SAN storage. As part of your business risk management and business continuity planning you need to define backup policies. You must be prepared to deal with disasters. Disasters can have physical, social, political, terrorist and many more origins with direct impact such as unavailability, loss of data or indirect rippling effects on the business (loss of market share, just to name one). They highlight the need for having a business continuity plan. You usually cannot afford to wait for a disaster to happen to start thinking about the way to deal with it. Dealing with data integrity and availability is one aspect in the overall business continuity plan to consider.

☒	Question	Why this matters
☐	6.1.10 Do you have clearly defined security policy and set up procedures?	Another aspect of business continuity management is information security. Information Security is about protecting a very valuable company asset: Information. It must be well protected. Information needs protection from unauthorized access and manipulations. We do this by carefully managing confidentiality, integrity and availability. Security audits are vital in determining security risks and defining business continuity plans. You should not only consider security tooling on the ICT Infrastructure (virus scanners, firewalls, etc.) but also look at social risks such as physical access to systems, networks, USB sticks, printouts etc.
☐	6.1.11 The Cloud is there to stay. What is your cloud strategy?	For simple workload such as mail, messaging, collaboration public cloud solutions such as Microsoft Office365 are becoming more and more established. Private cloud solutions are the way to go for setting up powerful and flexible architectures. This offers also the possibility for hybrid cloud solutions where part of the solution runs in the public cloud. The Microsoft Azure strategy allows this flexibility.

Chapter

7

Project Approach

All work expands to fill the time available

Cyril Northcote Parkinson (Naval Historian, 1909 – 1993)

Solution is one thing, Practices another - What makes an implementation successful?

S ELECTING the right product to meet your business requirements is one thing. Yes, choosing the right product to meet your business requirements helps. It ensures there is a good fit between what you want to achieve as a business and what the product has to offer in processes, functions and features.

But there is more… How is the solution implemented in the organization?

Analyses show that there is considerable risk that the customer expectations and the actual results do not

Figure 28 - How to implement the project?

match or even that the project becomes a complete failure. Even if the implementation lives up to the expectation there is a risk that the implementation takes longer and exceeds the budget.

Various elements make up a successful implementation. We call them the fabulous six. Here they are:

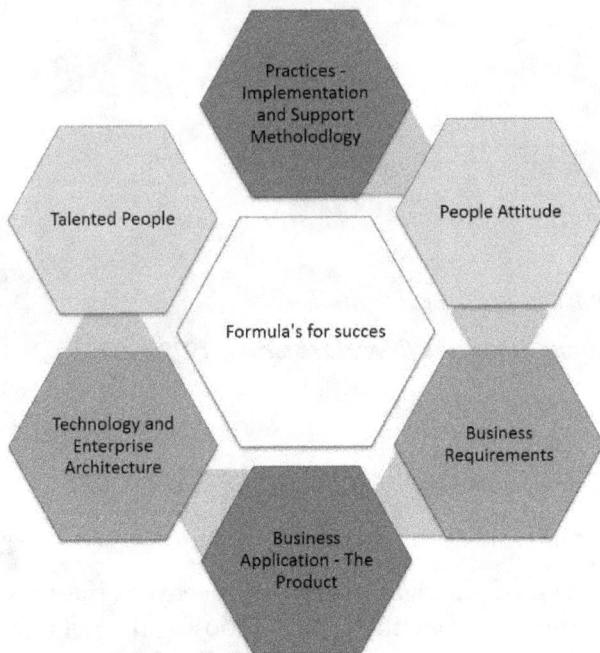

Figure 29 - The fabulous six

Technology and Enterprise Architecture

You need to be able to rely on the right technology. The MECOMS™ and Microsoft Dynamics AX are built on state-of-the-art technology. It is an integrated technology stack. So, no need to worry on how Database, Service Oriented Architecture, BI, User interface will work together. It is all there and it works.

Business Applications – The Product

In previous chapters we have seen that the energy and utility market really has specific business needs. The solution needs to be able to deal with these requirements. An excellent fitting application (standard product) is quintessential for successful implementations.

Business Requirements

If it is not clear what you want to achieve with the application, you cannot start a project. Even when using an iterative, agile approach there needs to be a view on where to go. We recommend using the business model canvas to get a good understanding of what the business wants to achieve and how they are going to do this. It goes without saying that business requirements are leading and need to be supported by the business application and technology.

Talented people

Any business, any implementation project success hugely depends on having the right, knowledgeable people in the organization. Yes, this is a scarce resource. But by putting the right people on the team an excellent pool of knowledge can be built both in business and in application implementation.

People Attitude

Yes – successful projects require excellent people with the right attitude. It is not only about having business and application knowledge. It requires the attitude to be successful. Many books and articles have been written on this topic. Research shows that projects where there is real team work and where people feel happy about their contribution to the implementation project are far more successful. Having a project steering committee with senior management from the utility and from the system integrator that motivates people can be very advantageous in the overall project success. If you do not have the people, the organization on your side, even the best product supported by the best implementation practices will fail.

Practices – Implementation and Support Methodology

Finally, there are the implementation practices. How are you going to handle the project? What are the processes you are going to manage? How are you going to govern the implementation project and manage the application once it is implemented? This is what this chapter is about.

The Sure Step Methodology

To ensure effective implementation of the business requirements one needs a proven, reliable and predictable methodology.

A methodology can be defined as follows:

> "A set of methods, practices, standard processes and concepts designed to ensure a repeatable, disciplined and manageable implementation of business requirements."

For the Microsoft Dynamics Business Applications, the Microsoft Dynamics Sure Step methodology has been developed. MECOMS™ offers extensions to the methodology to ensure that the right tools are available to implement MECOMS™ in a successful way to meet the business requirements. The MECOMS™ extensions are based on more than 20 year experience of having ISO-900x/TickIT certified processes.

Methodology Deliverables

The methodology offers you:

- Standard Processes
- Template documents – to support the processes and to prevent reinventing the wheel.
- Tools
- Role – responsibility matrix

The methodology is supported by a tool that delivers process descriptions, flows, responsibilities, tools and templates in a structured way either on the local machine or using Microsoft SharePoint, Intra- or Extra-net.

Key Components in the Methodology

Roles

The Microsoft Dynamics Sure Step Methodology groups the practices for the individual roles in a project and operational organization. These roles include, Project Management, Subject Matter Expert, Application Consultants, Development Consultants, Testers, Trainers, etc. and are related to both the customer organization and the consulting/system integration organization.

Project Phases

The Microsoft Sure Step Methodology covers the complete life cycle from solution selection to operations of the business application. There are 6 Project Phases:

- Diagnostics Phase
- Analysis Phase
- Design Phase
- Development Phase
- Deployment Phase
- Operations Phase

Cross Phase Activities

If you look across the project phases there are activities that cross the phase boundaries that are essential in the delivering the solution. Within the Microsoft Dynamics Sure Step and MECOMS™ these are called cross phase activities. There are 9 of them:

- Program Management
- Training
- Business Process Analysis
- Requirements and Configuration
- Custom Coding
- Quality and testing
- Infrastructure
- Integration and Interfaces
- Data Migration

Project Types

Depending on the organization's culture (Business Users, Project Staff, IT Staff, etc.) and the level of expertise in the project teams you can opt for a Waterfall approach or an iterative approach. The methodology supports various project types

- Standard
- Rapid – For fast track implementations of the standard product with no or very limited customizations.
- Enterprise – For the more complex implementations

- Upgrade
- Agile – Using an iterative approach typically is possible when talented Subject Mater Experts are available both at the customer and the consulting end. It requires a different project culture.

Benefits of the methodology

Benefits of using a standard methodology are numerous. We just briefly touch on some important benefits:

For the Utility Company

- A proven mechanism to deal with showing the business value of the implemented applications
- Agreed responsibilities by providing a repository of tasks and deliverables to clarify roles and responsibilities, on both the customer and consulting side and plan resources accordingly.
- Confidence in delivering a quality (let's define it the Crosby way – Quality is meeting the customer requirements) solution to budget, requirements and scope, planning (actually these are the customers' requirements).
- Ensuring the business application integrates in the company's overall IT landscape.
- Ensure there is excellent governance in place to deal with project issues. The earlier risks and issues are detected and listed the higher the chance that they do not impact the timeline (it would not be a project if no Project issues were coming up).
- Excellence in execution and hence optimizing the cost/benefits.
- A common language, Lingua Franca of terms across the project and business teams. Most utilities staff are in the business of running a utility and do not get often involved with implementation projects.
- Have a thorough view on project issues and risks allowing for swift decision taking.
- System Integrator independency

For the Implementation Consultant/System Integrator

- Best practice
- Showing the customer the business value of the application.
- Demonstrate a proven approach to the project.
- Tool to explain to the Utility company (and also within the own organization) how the implementation project is executed.
- A platform for knowledge management and continuous improvements to the methodology.
- And obviously also all the reasons that apply to utility. Successful projects are about team work between the utility and the System integrator. They should be in the same page!

Methodology support

The methodology is supported by a tool.

The MECOMS™ implementation methodology is available as an add-on to Microsoft Dynamics Sure Step that comes with the MECOMS™ application software. Templates, processes, documents, checklists and much more are available within the Microsoft Dynamics Sure Step tool.

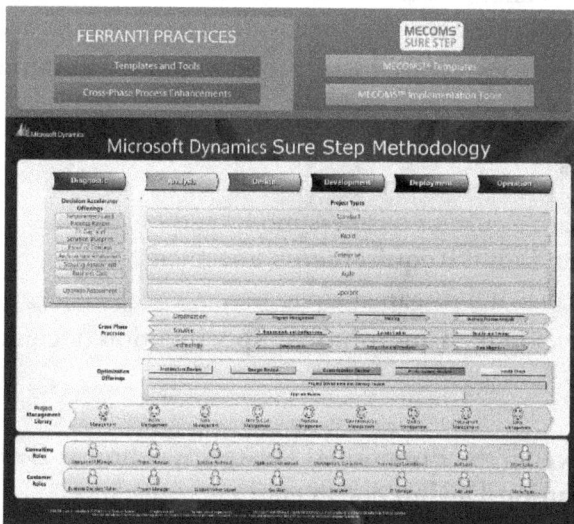

Figure 30 - Microsoft Sure step Implementation methodology

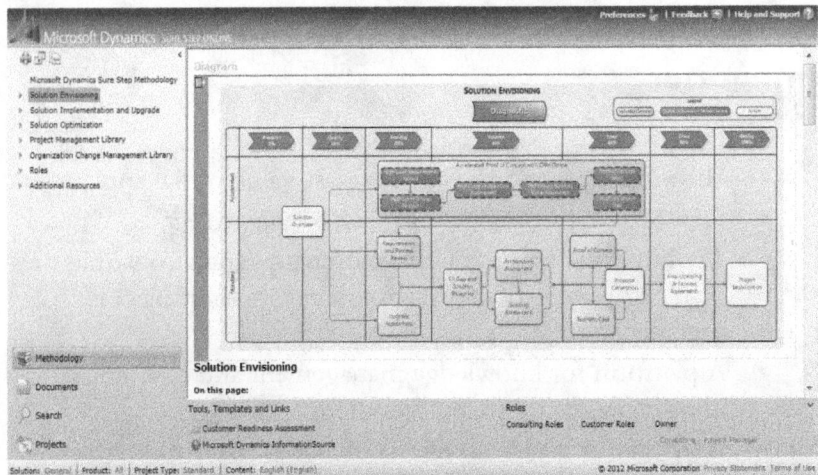

Figure 31 - Microsoft Dynamics Sure Step Online tool

Project Phases

A project is made up of a number of sequential phases. Each phase has a set of deliverables and exit conditions to ensure quality. Let's get a closer look at what each phase is about:

Diagnostics Phase

- Evaluate a customer's business processes and infrastructure
- Assist the customer with their due diligence cycle, including ascertaining requirements and their fit with the solution and assessing the resource needs for the solution delivery
- Prepare the project plan, proposal and the **Statement of Work**

Analysis

- Analyze current business model and finalize the **Functional Requirements** document
- Finalize the fit-gap analysis
- Develop the **Environment Specification** documentation

Design

- Develop the **Functional Design, Technical Design** and **Solution Design** documents
- Finalize the data migration design

- Establish test criteria

Development

- Finalize **configurations and setup** of the standard solution
- Develop and finalize the **custom code** that is required to support the solution
- Conduct functional and feature **testing** of the solution
- Solution testing including process, integration and data acceptance testing.
- Create the **user training documentation**

Deployment

- Complete End User Training
- Perform User Acceptance Testing
- Finalize Production Environment
- Perform final Data Migration
- Prepare Cutover to Production and finalize System Go-Live. Any subsequent environment changes are placed under the established system change control process

Operation

- Resolve pending issues
- Finalize user documentation and **knowledge transfer**
- Conduct a **project closure review**.
- Provide **on-going support** (activities that continue through any future involvement with the customer after the project is closed)

Project Cross Phase activities

In the previous paragraph we reviewed the phases of a Sure Step Project. There are 9 cross phase processes explaining step by step the tasks, deliverables and responsibilities for each of the above phases.

Organization Perspective

- Program Management / Project Management Library

- Training
- Business Process Analysis

Solution Perspectives

- Requirements and configuration
- Custom Coding
- Quality and testing

Technology Perspective

- Infrastructure
- Integration and interfaces
- Data Migration

☒	Question	Why this matters
☐	7.1.1 Does your project have a clear set of business requirements?	Embarking on a project without a clear scope and business objective is fatal. A project has a start time and end time and well defined business objectives to be met. This does not mean that there is no room for agility in a project but you need to deal with a drifting scope in a controlled way. Did your business requirements cover all desired business processes in scope? Clarifying the business processes as input for your business requirements will facilitate the fit / gap analysis against standard product and will help to determine the priority of your requirements. The set of requirements will determine whether your project was a success during the acceptance testing. During acceptance you will want to know if all requirements have been met. Last but not least, clearly outlined future business processes will help you to manage your organizational change. The Project Charter needs to define all of these elements before the project can start.

X	Question	Why this matters
☐	7.1.2 Did you decide on the type of project approach to use i.e. Waterfall or Agile?	You need to define the project approach and phasing in the project charter and determine how you are going to run the project. It needs to be absolutely clear to everyone how the result will be reached and what the deliverables are. Do this up-front to avoid having complicated, time consuming discussions during the project. Working in an agile way raises the bar for knowledge and experience of the subject matter experts, analysts, implementation consultants and developers. An agile, iterative approach can offer you greater visibility as you track the progress by looking at the actual software rather than relying on the progress reports and seeing the result at the end. Also there is more flexibility in determining how the business requirements are implemented. If Agile is used the right way the quality of the product should also be better. Whether you decide on using Agile, Waterfall or any other approach you'll always need to deal with the issue of managing the budget and planning and with business users that keep changing their mind. Also look at the availability of project resources to support the chosen project methodology.

☒	Question	Why this matters
☐	7.1.3 Has a project charter been setup describing the goal and approach?	Ensure you define the goal and success factors of your project. Too many projects run without an excellent project charter or are managed in another way than the charter describes. Ensure the charter describes, (project) organization, responsibility, governance, milestones, delivery milestones (and exit criteria for them), scope, risks and mitigations. It should be a document that is seen as the reference for both the utility and the system integrator.
☐	7.1.4 How are you dealing with gaps between the solution and your business requirements?	When implementing a standard solution there are bound to be gaps between the product solution that you are implementing and your business requirements. A good practice is to maintain a gap list. Not all gaps need to be filled. There might be workarounds possible.

☒	Question	Why this matters
☐	7.1.5 Have you decided on the test strategy and test planning?	How are you going to determine that the system you set up meets the business requirements? Testing is the answer. Therefore you need to come up with a test strategy. What are you going to test? When are you going to test it and on which environment? How will you track the progress of testing and issues? What are the acceptance criteria per test phase? Here is the trade-off. You cannot test everything. The cost of testing everything is just too high. At the same time, the cost of solving an issue that slipped through the testing cycle is higher than the cost of solving it during development. So you need to determine what the right level of detail is for the tests you want to run. A good rule of thumb is to bear in mind that processes with consumer impact probably need more test attention. Also think about how you are going to staff the testing both in terms of knowledge and timeline. Accept you cannot test everything! Typically these elements are included in a MECOMS™ test plan.

☒	Question	Why this matters
☐	7.1.6 Are test scripts available?	Do not base your testing strategy on monkey tests only. Develop test scenarios and scripts and use these as the basis for your testing. Think about the test data that you are going to use. Is the test data representative for the real live situation? Are you using the various types of instances in your test i.e. different types of customers, types of contracts, types of meters, types of connection, etc.?
☐	7.1.7 Does your data migration strategy allow for high quality data in the new system?	One of the most complex domains in setting up a new system is data migration from the existing systems to the new systems. Does the source and destination application allow for time slicing allowing for reconstruction of information in the past? MECOMS™ does. But main utilities do not have time sliced historical data. You need to come up with a strategy for data migration. You may need to decide to only migrate actual data for specific data elements. You will need to define exit criteria for a successful data migration run. Chances are that 100% automatic migration is not possible and that you will need to do some manual cleansing. More on this in a separate chapter!

☒	Question	Why this matters
☐	7.1.8 Are the exit criteria for each of the phases clearly defined?	Spend time while developing the Project Charter to define the exit criteria for each phase/activity of the implementation project. Who needs to sign off? What are the quality requirements for the deliverable? Get the right business and IT users involved. Chances are that your utility organization has to deal with a business / IT divide (not uncommon and not only in Energy and Utility market!) Deal with this in a professional way, use your social skills. Do not take short cuts! You will regret it later on in the project.
☐	7.1.9 Is there a clear communication strategy to the various parties involved?	The secret is "do as you say and say what you do" For the "do – as you say" part there is the project charter defining what, when and how. The other part is covered by project progress reports. Typically they are neither written in a language the various stakeholders would understand nor are they sufficiently focused on individual stakeholders. So, put your communication skills into practice. Think of developing a newsletter or Intranet postings and periodic briefings. Focus on the various impacted parties in the project. Plan time for celebrating achievements and achievers! Remember, the people make the project successful.

☒	Question	Why this matters
☐	7.1.10 Is a Cut over / Go live scenario defined?	Most systems are non-greenfield implementations. You need to define a plan on how the new system is taken live and the old system is switched off. What is the sequence of stopping processes in the old system and starting up processes in the new system? What is the plan of action with outstanding running scenarios in the old system? Do you move them to the new system? How do you deal with information you send and receive from other market parties in your utility landscape? How long is the data migration going to take? Is this acceptable to the business? Can you execute data extraction (from the old system) and load (in the new system) in a parallel way? What is the impact for the customer of the utility? Do we need communication to these customers? Can we work with incremental migration and thus migrate a set of customers at a time?

☒	Question	Why this matters
☐	7.1.11 What is the strategy for issue management?	Issues will come up in your project. Do not let them drift. Make one single inventory and make someone responsible for resolving the issue. Plan time to resolve issues and review outstanding issues. Create a culture where issues are flagged and logged as soon as possible. This gives you more time to resolve them. Do not be an ostrich! Do not put you head in the sand. Far too many projects handle issues in an ad-hoc way. This does not lead to a quality result.
☐	7.1.12 How is overall project progress monitored and reported?	Make no nonsense reports focusing on the headlines. The report should not only report the issues but also the corrective actions.

☒	Question	Why this matters
☐	7.1.13 What is the strategy for dealing with project risks?	You should make a risk register and determine mitigations for each of the risks. You could do a brainstorm with the team – Let each team member come up with 3-5 risks he sees for the project and stick them on an X-Y graph (x = Impact, Y = likelihood) Make someone responsible for dealing with each of the risks. Too many projects make a risk register in the project but have alpha-male project managers unwilling to review them during the project as they might show weaknesses in the organization. A good practice is to review this with the team on a regular basis. Use risk management during your decisions and phase sign-offs.

Project Approach

Chapter

8

Data Migration

The world as we have created it is a process of our thinking. It cannot be changed without changing our thinking.

Albert Einstein (physicist, 1879 – 1955)

Why this topic has a separate chapter?

WHY should we spend a separate chapter on Data Migration? From experience we know that this task should never be underestimated. Implementing a new solution requires it to be filled with data from one or more legacy systems. Data migration might also involve integration of data related to companies that have been acquired. In many cases, having the right data migration approach in place will make the difference between success and failure.

Data Migration might involve several source systems. The data in these systems can be inconsistent. To make things worse, you might be missing data to make your new systems work perfectly.

Figure 32 - The art of data migration

You will need to define exit criteria for successful data migration. No data migration can give you 100% data quality (if it would you would not be able to afford it).

Data migration process

Data migration is an art. You need to have the right business knowledge in the migration team to ensure the impact of decisions is well understood. Correct data is required to support the new business processes. It is not our ambition to describe the complete process. Consider:

- You need a clear understanding of the data structure of the source systems and the destination system. For MECOMS™ the destination data model is what we call the MECOMS™ conceptual data model.
- Determine ownership in case of data redundancy.
- You need to understand and determine which data need to be migrated. This is often called data profiling.
- Consider the value of old data for the company.
- Define the migration mapping rules between the source and destination systems
- Determine rules for conversion and cleansing of data.
- Develop efficient conversion scripts.
- Determine the acceptance criteria for the quality of data and execute your test scripts to determine the quality of the migrated data.
- Determine exit criteria for each of the process steps.
- Plan for migration dry runs to understand the process and quality. Ensure the right IT infrastructure is available.
- Determine how long data migration will run and the impact on the production environment.

Data migration is a complex theme. We refer to MECOMS™ E-learning material for further details.

☒	Question	Why this matters
☐	8.1.1 Have you planned for Business commitment to the Data Migration	Utility systems typically are implemented by Projects with an IT background. This is great as these skills are definitely required to ensure a successful implementation. The new system has a set of innovative functions. But to use these you need to have the right data available. Business needs to be very aware of the impact of not having quality data available. This is a call the IT staff cannot make as it impacts the way the business needs to use the system.
☐	8.1.2 Who has the final call in data migration decisions?	The data migration project should define clear responsibility on decisions regarding data quality, availability and migration rules. An implementation cannot afford to have a deterministic behavior as a result of taking the wrong data migration rules.
☐	8.1.3 Have you defined your migration strategy?	Various approaches can be used for data migration. One could go for a Big Bang Migration where all the data is migrated in one go. Alternatively one could go for incremental migration. Sometimes the number of records or complexity of the migration rules is such that manual migration is a better option.
☐	8.1.4 Do you have KPI's in place to monitor data quality?	Regardless of the effort in the data migration stream, achieving 100% correct data is not achievable. Therefore it is essential to determine KPI's that allow you to monitor the quantitative and qualitative aspects of the migration.

☒	Question	Why this matters
☐	8.1.5 Do you have Data Migration Rules in place?	Each field in each record in the new system needs to be populated with data from existing source systems. Automated migration rules will help optimizing the migration process.
☐	8.1.6 Are Data Quality Rules defined?	It is good practice to measure the quality of source data and target data. This gives an insight in the corrections that are required. Preferably these are executed automatically. This also helps in developing KPI's to monitor progress.
☐	8.1.7 Have you defined exit criteria for a successful data migration?	Using your KPI's, you can define criteria that must be met before going live with the new system. It is essential to make this decision a mathematical one instead of an emotional one. Often in time critical projects, data quality is sacrificed against time. There is nothing wrong with this as long as it is a conscious decision. Cleaning data in a live situation can be an expensive proposition.
☐	8.1.8 How are you going to deal with historical data?	Migration of the current situation is the easy part. If you want to migrate the history this can be more difficult. This relates to the availability in the source system of time sliced information.
☒	8.1.9 What is the strategy for outstanding scenarios	In live systems process scenarios are running. The process scenarios could run days or even weeks. An example could be a supplier switch, customer switch. How are you going to bring the outstanding scenarios to the new system? Can you let all outstanding scenarios run to an end? Or do you want to migrate them to the new system?

☒	Question	Why this matters
☐	8.1.10 How are you going to retire legacy systems?	Determine a strategy to switch off old systems. You need to consider what you want to with old data.
☐	8.1.11 Do you have a migration time line in place? And what about your optimization strategy for the time required to run the migration?	The extraction, cleansing and loading of data into the new system takes time. Once the "go" is given timing is of the essence. Dry run migration on the appropriate hardware will give you an insight how long it takes to complete migration. Plan for a number of dry runs during the project to optimize the time it takes to do the migration. Also all activities from shutting down scenarios, systems, migration and slow start-up, to full go-live should be defined. Ensure you have the right specialists available for each phase.
☐	8.1.12 What is your strategy if the data migration fails?	You should develop a fallback plan in case the data migration for go-live fails. This is triggered when any of the exit criteria of the data migration time line are not met. The plan should take the old and the new system into account.
☐	8.1.13 Do you have a Data Freeze strategy?	In live systems data continuously changes. For data migration you must freeze or snapshot the data. This data is then used for the migration process.

Data Migration

Chapter

9

Application Management

Love your tools and they will love you.

Adrian Jones (A great golf teacher)

Why Application Maintenance

YOU have worked hard to go-live with a new application on time and within budget. There are no blocking issues left. It is time to have a party.

Setting up the application takes you through a structured life cycle. You need to consider how to deal with support, enhancements and optimizations.

Figure 33 - The art of application maintenance

You need to think ahead to avoid falling into the traps that all support teams are faced with. Just a few examples – No structured access to professional support, critical application with difficulties to retain staff and therefore knowledge, documentation not up-to-date, alignment missing between infrastructure oriented support staff

and business application support staff, database administration function not aligned with business needs. Another typical one is that the lead time to implement changes whether they are technical, business or regulatory based, is just too long.

You need an application management strategy to define processes and expectations for your users.

Microsoft Operation Framework, MOF and ITIL

Microsoft created the Microsoft Operation Framework or MOF for short as a guideline for IT professionals to implement reliable cost effective IT Services. The goal of MOF is to support the entire IT life cycle. So how does it relate to the Sure Step methodology presented in the previous chapters? In fact, the Microsoft Sure Step methodology and the MECOMS™ extension focus on business application configuration and setup whereas MOF is more an overall methodology to plan, deliver, operate and manage IT solutions.

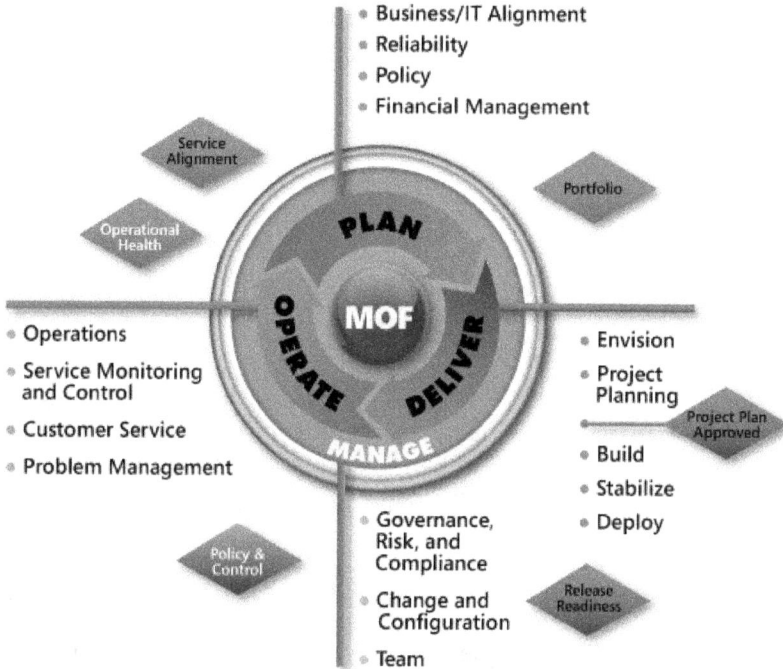

- Business/IT Alignment
- Reliability
- Policy
- Financial Management

PLAN

Service Alignment

Operational Health

Portfolio

MOF

- Operations
- Service Monitoring and Control
- Customer Service
- Problem Management

- Envision
- Project Planning

Project Plan Approved

- Build
- Stabilize
- Deploy

- Governance, Risk, and Compliance
- Change and Configuration
- Team

Policy & Control

Release Readiness

Figure 34 - Microsoft Operations Framework

The MOF framework identifies a number of phases. If you read through the following phase definitions, you get to grasp the difference with the Microsoft Sure Step Methodology.

The **Plan Phase** provides guidance on how to plan for and optimize an IT service strategy—it helps to deliver services that are valuable and compelling for the organization, predictable and reliable, policy-compliant, cost-effective and adaptable to changing business needs.

The **Deliver Phase** helps IT professionals to more effectively deliver IT services, infrastructure projects or packaged product deployments and ensures that those services are envisioned, planned, built, stabilized and deployed in line with business requirements and the customer's specifications.

The **Operate Phase** helps IT professionals efficiently operate, monitor and support deployed services in line with agreed-to service level agreement (SLA) targets.

The **Manage Layer** establishes an integrated approach to IT service management activities. This integration is enhanced through the establishment of decision-making processes and the use of risk management, change management and controls.

If you are familiar with the ITIL (IT Infrastructure Library) V3 Framework, you might think, so what... Indeed the Microsoft Operations Framework (MOF) is based on the same principles as ITIL. Both MOF and ITIL are IT Service Management frameworks (ITSM). Actually MOF evolved from ITIL V2. These frameworks are set up to transition IT departments focused on IT technology to IT services covering people, processes and technology. The reason this is important is that people and processes make up for 80% of IT services downtime on average worldwide. If you want to increase up time and align the IT service to the business users you had better start setting up a framework. The cloud paradigm makes IT departments realize that they need to think about services.

Service Management – Service Desk processes

The service desk is the single point of contact for questions, issues, incidents or problems. We will not elaborate on the differences between MOF and ITIL. The goal is to have a structured process to allow business user to report issues (we are not going to be fussy about terminology) More importantly the service desk function needs to make sure it gives feedback to its customers. If this does not happen in a structured way, users get frustrated and look for alternative ways to get things done.

Service desk call intake can be implemented with physical call center operators or self-service functionality. Microsoft System Center Service Manager is an application that can be used to automate the process and provide reporting and KPI functionalities on this process. The Microsoft System Center suite not only offers service management but also monitoring and operations management functionalities.

Some calls to the service desk can be handed by the first line others are forwarded to second and third line support.

The ITSM framework also imposes requirements KPI's on processes. These should be IT-customer oriented rather than IT infrastructure oriented. What's the point, we can say "the network was up" if an important service dealing with meter reading was down resulting in the meter process to fail.

Knowledge management

One of the pains in supporting and running a solution is having the knowledge to do so. There is a constant flow of information. The overload of information makes knowledge management essential. What is your onboarding process when hiring new staff? Organizations benefit from becoming learning organization by making learning routine. By doing so knowledge management becomes a driver for innovation and cultural change and feeds decision-making. As for all chapters in this book, there is so much more to say about the topics but the space is limited. We hope, we can encourage you to develop a knowledge management strategy for the organization.

Exciting technologies such as Microsoft SharePoint, Microsoft Yammer can help you to set up a collaborative environment for your organization by using:

- WIKI libraries – extending the MECOMS™ process wheel – to have process descriptions.
- Process libraries – extending the MECOMS™ process wheel – to store process flows.
- Discussion forums/Yammer feeds as a structured platform to promote discussions on the business and IT implementations
- Collaborative workspace – to work on projects, changes, etc.
- Instant messaging

With Microsoft Office365 you could even consider running this in the cloud.

Business Processes and IT infrastructure

Capturing this knowledge can be daunting. You need to get a good insight in Business Process, IT services and IT Components.

The relation is as follows:

- A Business Process uses a number of Services
- A Service uses a number of IT Infrastructure components
- The IT infrastructure components can be hardware and have some redundancy

Business Processes - Functions	Data base	SDAS Service	FTP	User AOS	Batch AOS	IAOS	Network
Receiving Meter Reads	√	√	√	√			√
Validating Meter Reads	√			√			√
Billing	√			√	√		√
Call Centre	√			√			√
Invoicing	√			√	√	√	√
Market Interfacing	√		√	√		√	√
Credit Management	√			√			√

Process use Services
Services impact processes
Infra impact Services
Services rely on infra

Physical IT Infrastructure	Data base	SDAS Service	FTP	User AOS	Batch AOS	IAOS	Network
LAN	√	√	√	√	√	√	√
WAN			√			√	
Database Srv	√	√		√	√		
User Application Object Svr				√			
Batch Application Object Sr.					√		
Integration Application Object Sr						√	
Smart Data Application Server		√					

Figure 35 - The BSC Matrix

This you can visualize in Business Process, IT Service, IT Component matrix or BSC matrix for short. It shows you on which services, business process rely. It shows you which infrastructure components each service uses. This helps when building a disaster recovery strategy as well

We could have included this also in the architecture- or the implementation methodology chapter. Indeed, this is very useful in the implementation project phase as well.

Change Management

Yes, changes will occur. Business changes, regulations change and technology changes all result in inevitable MECOMS™ business application changes.

Sources of changes include:

- Software patches from core technologies
- Changes to business process to improve efficiency
- Changes as a result of legislations

- Software bugs
- New versions of software components.

The goal of the change management process is to ensure:

- A good trace of the change request issued in the organization is available.
- An assessment of the impact of the change.
- Cost/benefit analysis and risk analysis for changes is worked out.
- Set priorities as there are typically more change requests than resource/budget available to implement them.
- Approval is given to go for changes.
- The implementation of the change request is monitored and reviewed.

Change advisory board is typically the authority that takes the decision in determining the change process.

Not having a change management process leads to chaotic cost management, frustrations and unhappy users.

Release Management.

Change requests, software defects and software updates result in changes in the software code base. The Requests For Change (RFC) that get implemented are governed by the change management process. The question thus becomes when and how are we going to implement them? This results in the need for a role that takes this up. Usually this is called the release manager. The release manager should:

- Plan the release – What is included in a release based on priorities and available resources.
- Determine the number of releases in function of time.
- Ensure releases are properly tested.
- Plan for installation of the release and the impact on running business applications. Considering down time.
- Serve as facilitator and liaison officer between business and IT.
- Have an excellent troubleshooting mind.
- Be a great influencer and coordinator to get things done by developers and to manage the expectation of the business.

- Determine risks of any release.

The release management role is an essential role as it brings together, the planning, management skills of the project managers, the design and architecture skills of the architect and business reasoning and influencing of the business reason with a clear focus on risk management.

The risk management aspect is extremely important as you do not want to have unnecessary downtime of the application.

The more agile you work, the more releases you have. You still need to get the testing and sign-off in place. There is continuous tension between the desire for stability and the pressure to implement changes fast. If you want to work agile, the availability of people with business knowledge for specifying the requirements, testing and acceptance should not be underestimated.

Having a standard schedule of periodic releases (once a month, once a quarter,…) and hotfix releases usually meets the needs.

Business Continuity Management

Disasters do happen. Disasters are not something that only happens to others. Various frameworks are set up to classify disasters. We give you a framework that we like to use and that can be extended as required:

- Nature – floods, hurricanes, lightning, fire
- Internal to company
  - Planned
    - Sabotage
    - Malicious Data manipulation
    - Strike
  - Accidently
    - ICT infrastructure breakdown
    - Loss of data
    - Loss of knowledge due to staff churn.
    - Business failure
- External to company
  - Security intrusion

Perform a threat analysis and define the probability of happening for each event. For each event (grouped in some classification), an impact and corrective action plan can be identified. The impact can be determined in financial terms. Once you have this information you can start defining your business continuity plan. Some risks can be mitigated by insurance coverage while others sometimes require substantial investments in business continuity measures. Due to low probability for some of the events, it is sometimes difficult to have a solid business justification for investment in business continuity measures. Think of the Fukushima disaster – What was the probability that a tsunami would hit and what was the estimated impact?

To summarize, business continuity is complex stuff. Give it consideration in defining the projects and running business applications. Some precautions can be taken without a huge cost impact.

Information Security Management

A special area of attention in the business continuity management list is Information Security Management. In principle the mechanism is the same. You identify the threats to the ICT assets; you identify the vulnerabilities and determine the impact.

Also for this area, standards have been defined to help organizations to set up appropriate policies and procedures to manage the risks. ISO 27001 is such a standard. The standard builds on the Deming continuous improvement cycle 'Plan-Do-Check-Act'. If your organization sets up an Information Security Management System it can then be audited and certified by external accredited organizations against the standard.

Larger organizations usually have an information security manager. Regardless of the size of your organization, consideration should be given to the security aspects of the solution.

Why spend a separate paragraph on security management, it is part of business continuity management? Indeed! Security is not a purely technical ICT technology issue though. Security is to a large extent depending on people and not only on technology.

Having said that, consideration should be given to obvious things

- Firewalls are included in the design
- Virus/worm detection needs to be in place

- There is a patch implementation policy for the core infrastructure
- An intrusion detection strategy is available
- Internal and external user administration process must be excellent.
- Promoting security awareness to company staff is an essential part of building any security or business continuity strategy.

☒	Question	Why this matters
☐	9.1.1 Have you developed a Business Continuity Plan?	Your business will be faced with disaster, large scale (hopefully not a lot) and small scale ones. Your business needs to prepare for this! Make an inventory of threats, vulnerabilities and the impact if disaster hits? For each of the events you should develop a plan of attack. By doing so you will know what to do when disaster strikes. Security management is one aspect of Business Continuity Management. You should consider ICT infrastructure disaster (server, network, disk failure, do not forget software failure)but equally if not more important are the people aspects and external factors. The objective is to deliver a business service with a Service Level Agreement.
☐	9.1.2 How do you deal with Database Administration?	Running a business application requires you loving the data in your system. This is the task of the database administrator. This is a technical role in terms of configuring, maintaining, administrating and monitoring the database management software. But, it also requires business insight! Consider the Database Administrators (DBA) Role in allocating and planning future storage based on the business inputs (these are not Gigabytes figures but expected business transitions), managing evolutions in the data schema, optimizing the performance of the database, planning of backup and recovery of the database, dealing with archived data, managing user access, etc.

☒	Question	Why this matters
☐	9.1.3 What is your policy and process for Security Auditing?	Utilities deal with information. This information is a key asset for the company. The fact that a solution like MECOMS™ is based on a well-designed, integrated, proven technology stack helps to reduce information security risks. But, this should not stop you from having a security audit plan to check for security holes, missing patches on operating systems and other core infrastructure. The security should also consider aspects such as social security, access to the office, waste printing paper i.e. the non ICT technology aspects.
☐	9.1.4 Do you have a good policy and process for user administration?	A solution like MECOMS™ for CIS and MDM has various users: your own business users, consumers and partners accessing the solutions, external system. You need to determine who has access to which information and functionalities.
☐	9.1.5 Do you have a skilled and well sized service desk for the business users?	New entrants in a market tend to under develop this aspect. At large utilities this is probably in place. The objective is to have a cost effective, efficient, business – IT aligned service to the user. Just like a consumer wants to get the answers from the Customer Interaction (CRM) help desk in MECOMS™ functional architecture, the business users want to get the answers to their questions.
☐	9.1.6 Do you have good reporting on service management activity in place?	Define the KPI's you want to use to drive the service desk function. This gives you a tool to take business decisions and motivates the service desk and the users. Good KPI's might include call aging, call intake and outstanding calls.

☒	Question	Why this matters
☐	9.1.7 What are second line and third line provisions?	If all calls to the service desk get resolved by the 3th line, something is wrong. There should be a good deal of knowledge available at the first line to solve issues. This will increase satisfaction of the people that are contacting the service desk. As the system matures the complexity of the problems raised increases. You need to have the right second and third line in place to support the very complex issues.
☐	9.1.8 What is your approach towards Knowledge Management? How does it fit in your overall business strategy?	Ensuring that the business strategy, the resulting business processes and the business application usage and architecture is well understood by the organization can make the difference between success and failure for a utility company. It is a determinant factor in ensuring whether the utility has the right Cost-to-Serve. You should consider building a knowledge management environment that allows for easy collaboration, clear process descriptions, e-learning. Creating e-learning can be a daunting task. In fact it can also be easy. Take a tool (http://www.articulate.com/ for example) and record what domain experts have to say about it. Forget that you can do a better job. Just do it. You would be amazed. It does neither take large budgets nor does it take a lot of time. All it is takes is a focused plan to make it happen. For critical resources, these are people with knowledge and skills that are really critical and not shared, a plan needs be developed to share the knowledge.

☒	Question	Why this matters
☐	9.1.9 How do you categorize changes? How do you determine which changes to consider?	Changes come in, if all is well, using a Request For Change template. They are either a result of service desk activity, business strategy and legislation discussion, patch strategies and IT landscape updating exercises. You need a structured process to clearly communicate what can be expected in terms of changes. It would be best practice, to set up a change advisory board which determines and manages the priorities and budget for changes. Rest assured there will always be more change than the budget you have available. Carefully selecting changes with maximum business benefit does the trick! This needs to be complemented by excellent governance and communication to ensure a successful change management strategy.
☐	9.1.10 Do you have a process to plan a release?	After the go-live a natural reflex might be to implement changes as soon as the software patch is available. You will soon discover this leads to a chaotic implementation of changes with no time to test the interdependency of changes. Build "peace and calm" in the organization and go for a periodic release. People need to know what to expect from support staff. This does not mean that you cannot immediately release a hot-fix for blocking issues. This should be the exception rather than the rule. One thing is sure, you need a release process.

☒	Question	Why this matters
☐	9.1.11 What is your strategic plan for onboarding standard technology and business application version updates?	MECOMS™ has the enormous advantage it is built on a technology stack that offers you rich functionalities without having to worry about the technology compatibility and integration. The underlying technology evolves; it gets better, richer, securer and more easy to use. You need to decide what your strategy is to grow with these evolutions. You need a patch strategy to determine which patches you will implement. You also need a strategy for the more significant changes such as new operating system versions, major release of the Microsoft Dynamics AX stack and the MECOMS™ application.
☐	9.1.12 What is your testing process before a release is installed in production?	Typically a release contains a number of changes. Some changes may not be ready to make it for the release. Sure, there will be interdependencies between individual changes. That is why it is important to run integral tests of functionalities before the release gets installed on the production environment.
☐	9.1.13 What is your release frequency?	You should try to find the optimum in release frequency. This very much depends on the organization. Having a weekly release optimizes the reactiveness to the business users but it undermines implementation and development times. Therefore it is much better to come with a reasonable schedule that also includes sign-offs of requirements and testing by business users.

☒	Question	Why this matters
☐	9.1.14 How do you communicate to the system users?	There should be clear communication on the release strategy to the business users. Having frequent communications encourages involvement.

Chapter

10

People and Organization

Here must all distrust be left behind; all cowardice must be ended.

Dante Alighieri, La divina commedia, Inferno (Poet, 1265 – 1321)

The story of the pig, the chicken and the rooster

L ET'S start this chapter with a story!

Once upon a time, there was a pig and a chicken living happily on a farm. The farmer took good care of them and the pig and chicken really appreciated that he gave them their favorite foods and made their homes clean and comfortable.

One day, the chicken was chatting to the pig and said, "Next week, it is the farmers birthday" "Great", the pig said, "I didn't realize, lets to do something for him". There was also a rooster nearby and he shouted " Yeeeeee"

All of a sudden the chicken said "I have a great idea, let's make an omelet with bacon, I know he loves it. Just think of how he will enjoy this". The rooster cried: "Excellent!!!"

The pig got quite anxious and nervous at that point it did not know what to say and wanted to do something really nice for the farmer. The Chicken shouted: "Yes, a great idea" and the Rooster responded "Yes!!!"

Finally the Pig responded: "For you this is involvement, you only have to lay two eggs but for me it is definitively commitment"

So, you have to start a project that dramatically changes and improves how you are our doing business. Yes, you know the product you want to implement. Yes, you know how to bring real excellence to the business processes of the company. Yes, you have organized for the best implementation methodology that offers you a structured approach to success.

There is only one part missing – the people that are going to make your project a success. This is often the most difficult part of the Jigsaw.

Pigs are totally committed to the project and accountable for its outcome.

Chickens consult on the project and are informed of its progress.

Figure 36 - One Team, One Strategy

Roosters can be defined as persons who strut around offering uninformed, unhelpful opinions.

If you want to make your implementation a success you need pigs (the pig must obviously die in order to provide bacon) and chickens. The chickens you will need to keep informed. You do not need roosters they will only complicate matters.

If you want to setup a successful implementation organization you will need to make sure you have sufficient pigs that are empowered to do their job in return for committing to and taking accountability for it.

As an organization you need to motivate a team of people from different departments and even different organizations:

- Business
- IT
- Business Consultants
- IT Consultants
- Supplier Subject Matter Specialists

An efficient implementation requires perfect orchestration of all people working on it.

Actually the pig, chicken and rooster story also applies to the business people in the organization. Think about it!

This is the reason why we spend a chapter just on "people". One chapter cannot give you in-depth insight in the complexity of leadership, management and psychology of individuals and groups but we hope it gives you some food for thought and triggers you to read more on this subject.

The analogy of the pig, chicken and rooster is a rather simple one. In this Chapter we will further discuss a more complete framework to understand who you have in the team and how they can effectively work on the individual tasks to make up your implementation.

Before we complete this chapter and jump to the Smart Questions, we want to give you a framework that will help you to identify sick organizations. This knowledge will assist you when dealing with the complexity of organizations.

Yes, there are plenty of other things to consider such as project team organization, executive sponsorship, team reporting, team feel, team celebrations. We cannot cover this all in this book but will give you some smart questions in the last part of this chapter.

The people factor – How to influence them?

Dealing with change in an organization is about getting the engagement and commitment of people to achieve the goals that are set out. Utility companies are under tremendous pressure to implement changes in processes and systems in a fast and efficient way. This is often the key to success for the company.

People are the key to meet goals and to be successful. As a leader, manager, Subject Matter Expert, Project Manager, Implementation Consultant in the utility or system integrator organization. How do you influence your stakeholder's audience? You should strive to create a collaborative context with a focus on objectives and goals influencing the stakeholders and cultivating innovative ideas, opinions and actions. If you want to meet the goal set for the project, this influencing is not limited to the typical hierarchical relations in an organization and people that report to you. You need to think about how you can actively influence people and structure your influencing strategy to meet your goals fast and efficiently. This should not be manipulative.

Research shows that there are several influencing styles and that the successful people use the right style at the right moment to meet their goals Some excellent work has been done in identifying influencing styles. Influencing styles include: rationalizing (use expertise, logic), asserting (rely on authority, personal confidence), negotiating (compromises, negotiation), inspiring (inspire and use metaphors) and bridging (influence outcomes by uniting or connecting with others).

Understanding Peoples Style

When you build an organization or build a project team you need to worry about getting complementary people on board to achieve the objectives. Yes, you learnt in the previous section about pigs, chicken and roosters. You could look at it in a more structured and less cartoonist way. Ichak Kalderon Adizes has written great books on leadership and human interaction. He models people in four categories:

Producer (P)

Producers like to produce. They stay busy and get things done. They focus on what we are doing and this makes the organization functional. They would rather work than go to meetings. They have no time for filing or planning. They finish one project and are ready to start another. They would rather work alone because they have no time to train others. Need something done? Give it to a Producer.

Administrator (A)

Administrators like to organize. They like rules, systems and procedures. They focus on how we do our work. If you do not have a policy, they will create one for you. They make sure we are doing things right and by the book. Time is a primary orientation. They are on schedule and would like others to be also. Need something organized? Give it to an Administrator.

Entrepreneur (E)

Entrepreneurs are thinkers and risk takers. They create and develop ideas on what to produce in the future. They are energetic and enthusiastic. They are always on the move, sometimes not looking where they are going. Need to start something new? Give it to an Entrepreneur.

Integrator (I)

Integrators bring people together and help people feel involved. Their focus is on who is helping and how well we are working together. This makes an organization "organic," or interdependent, like a living entity. Need to connect with others or resolve conflicts? Call an Integrator.

If you want to get works done, if you need to build an organization, it definitely helps if you understand the type of people you work with and how they complement one other to success.

Sick Organizations

People are social creatures requiring social interaction. The social interaction gives them feedback and feeling on their value. Under normal circumstances, emotionally mature people get self-esteem from their day-to-day activities and from trusted relationships with other people. We tend to call them self-motivated people.

The reality though can be different. Individuals or organizations can be dysfunctional or even neurotic. Most people in healthy organizations do not let whatever neurotic tendencies they have influence their day-to-day performance.

The reality is also that you will need to deal with people with different behavior to make the implementation project a success.

This is the complex field of organizational research. Many books have been written on psychoanalytic and psychotherapeutic aspects of team behaviors.

Manfred Kets De Vries has written some excellent books on this topic and identifies 5 neurotic styles that define a neurotic leader and can impact an organization. Obviously this applies to every member in an organization. We like to use this model to look at the effectiveness of an implementation and change project of solution.

Paranoid

Characteristics	Suspicion and mistrust of others; hypersensitivity and hyper-alter-ness; readiness to counter perceived threats; over-concern with hidden motives and special meanings; intense attention span ; cold, rational, unemotional
Fantasy	I really cannot trust anybody. A menacing superior force exists which is out to get me.
Danger	Distortion of reality due to preoccupation with confirmation of suspicions. Loss of capacity for spontaneous action because of defensive attitude

Compulsive

Characteristics	Perfectionistic; Pre-occupation with trivial details; insistence that others submit to own way of doing things; relationships seen in terms of dominance and submission; lack of spontaneity; inability to relax; meticulousness; dogmatism; obstinacy
Fantasy	I do not want to be at the mercy of events. I have to master and control all the things affecting me.
Danger	Inward orientation. Indecisiveness and postponement; avoidance due to fear of making mistakes. Inability to deviate from planned activity. Excessive reliance on rules and regulations. Difficulties in seeing the "big picture"

Dramatic

Characteristics	Self-dramatization; Excessive expression of emotion; incessant drawing of attention to self; a carving for activity and excitement; incapacity for concentration or sharply focused attention.
Fantasy	I want to get attention from and impress the people who
Danger	Superficiality; suggestibility. The risk of operating in a nonfactual world--action based 'hunches.' Over-reaction to minor events.

Depressive

Characteristics	Feelings of guilt, worthlessness, self-reproach, inadequacy. Sense of helplessness and hopelessness – of being at the mercy of events. Diminished ability to think clearly, loss of interest and motivation; inability to experience pleasure.
Fantasy	It is hopeless to change the course of events in my live. I am just not good enough.
Danger	Overly pessimistic outlook. Difficulties in concentration and performance. Inhibition of action and indecisiveness.

Schizoid

Characteristics	Detachment, non-involvement withdrawal; sense of estrangement; lack of excitement or enthusiasm; indifference to praise or criticism; lack of interest in present or future; appearance cold, unemotional.
Fantasy	The world of reality does not offer any satisfaction to me. All my interactions with others will eventually fail and cause harm so it is safer to remain distant.
Danger	Emotional isolation results in frustration of dependency needs of others. Bewilderment and aggressiveness may be the consequence

You might think why this chapter is in this book. This is a terrain that is much more difficult to explore for most business or IT people. Rest assured, the people make the difference when you build a business or project organization.

☒	Question	Why this matters
☐	10.1.1 Have you thought about staffing your project taking into account not only the technical skills but also to the soft skills?	To make an implementation a success you need the right technical skills. These include knowing the technology, understanding the business processes and being equipped with the right tools and methodology. But if an individual cannot work to a common set of objectives and goals, the project fails. It is the responsibility of the Executive Management Steering Committee and the project management to ensure an empowered team is carefully put together with the right skills and team feel that understands what needs to be done and how it is done. You will need pigs, chickens and hopefully no roosters.
☐	10.1.2 Do you have the right person on the right job?	People need to feel happy about their works and their achievements. Think about hard skills and soft skills of the individuals and imagine how they will work as a team.
☐	10.1.3 Do you have the right stakeholders on the project?	Make sure you have the right stakeholders on the project such as business stakeholders (Business managers and end users) and technical stakeholders (IT management and ITs operations staff). Remember Post-Go-Live these people will use the solutions so you need their buy in. Rather than just interviewing people to understand the requirements, actively engage them as participants in the team.

☒	Question	Why this matters
☐	10.1.4 Will the team work as an empowered organization or will it turn into a sick project organization?	Many projects and in fact organizations fail because of dysfunctional and neurotic problems. There should be trust in the team. This does not mean there cannot be control. Empowering people makes a lot of difference. The ambitious and achievement-oriented Generation Y, people born in mid-80's and later, is specifically known to change jobs if they are not kept informed, appreciated or if they do not receive reassurance.
☐	10.1.5 Are people available and accessible?	Many plans fail because they rely on people that are not available as they are doing other things. This might frustrate the people involved. It makes the support service project go in overspend and fail to meet the time constraints. If you get people to do a job, make sure they are available. Get their buy in so they can finish the job in time.
☐	10.1.6 Is the Staffing aligned with what needs to be done?	The project plan should have work breakdown structures with estimates for the work to be done. So it should be pretty simple to plan the work? Yes and no. You will need to ensure that the hard skills (practice, technology, business process, etc.) match the requirement of the job (and it might be the match is not 100%) You also need to ensure the soft skills match and that the inter-people-relations work out.

☒	Question	Why this matters
☐	10.1.7 Do you have the appropriate team communication in place and have you planned for celebrating success?	People like to know what is happening. Think about the success of the tabloids. People like to be acknowledged for their achievements. Create a project newsletter or support newsletter depending on the project phase. Bake a cake to celebrate a milestone or an achievement. Make time to celebrate an achievement and do not assume it is normal. It makes a big difference in building empowered teams.

Chapter

11

Some Case Studies

The interest I have to believe a thing is no proof that such a thing exists.

Voltaire (French philosopher, 1694 – 1778)

Where is MECOMS™ used?

ECOMS™ is designed as described in previous chapters, to deal with various commodities, markets and market roles. The solution has been developed with this essential design criterion in mind. This book should have given you a comfortable feeling that the MECOMS™ Solution is fit for the utilities market.

However, the proof of the pudding is still in the eating.

So how is MECOMS™ used and how does it bring value to the utilities business?

Check out the **on-line** Customer Stories page of the MECOMS™ website. Here you will find case studies and video testimonials

http://www.mecoms.com/customer_stories/

Chapter

12

Final Word

The saddest summary of a life contains three descriptions: could have, might have and should have.

Louis E. Boone (Academic Author, 1941 –)

HOW to summarize this book? There is so much more to say. The utility sector is undergoing tremendous change. Existing legacy IT solutions cannot keep up with the change in the market. Flexibility is required to support the new market models needed to support Deregulation, franchising and new challenges of distributed production.

Utility companies need cost efficient solutions for customer care and billing (call it CIS – Customer Information Systems) and meter data management. MECOMS™ powered by the Microsoft Dynamics AX flagship ERP product is such a solution.

The MECOMS™ information, functional and process models offer a vertical solution for Electricity, Gas, Water and Thermal Utilities (or mixed utilities) in every market role (integrated utility, grid operator, retailer, generators...)

The business solution is one thing. You need a powerful Enterprise architecture that grows with the application. Microsoft Dynamics AX and MECOMS™ offer a robust proven architecture.

You also need to consider implementation and application maintenance practices as well as the people-factor to make the solution a success.

All building blocks are here – Time is now to move to a new era of flexible and efficient Microsoft Dynamics – MECOMS™ powered solutions that meet the needs of energy and utility companies.

List of Figures

Bibliography

Some recommended inspiring reading used to develop ideas in this book:

- The pyramid principle, Logic in writing and thinking – Barbara Minto – Prentice Hall
- It Sounded Good When We Started, A Project manager's Guide to working with people on projects - Dwayene Phillips, Roy O'Bryan – Wiley
- Business Model Generation – Alexander Osterwalder, Yves Pigneur – Wiley.
- Energy Risks - Dragana Pilipovic – Valuing and managing Energy Derivaties – Mac Graw Hill.
- The Other side of Innovation – Vija Govindarajan, Chris Trimble – Hardvard Business Review Press.
- A Guide to Putting the Learning Organization to Work – David Garvin – Harvard Business Review Press
- Management/Mismanagement Styles – Ichak Kalderon Adizes Ph.D – The Adizes Institute
- Reflections on groups and organisations – Manfred Kets De Vries – Wiley

Getting Involved

The Smart Questions community

There may be questions that we should have asked but didn't. Or specific questions which may be relevant to your situation, but not everyone in general. Go to the website for the book and post the questions. You never know, they may make it into the next edition of the book. That is a key part of the Smart Questions Philosophy.

Send us your feedback

We love feedback. We prefer great reviews, but we'll accept anything that helps take the ideas further. We welcome your comments on this book.

We'd prefer email, as it's easy to answer and saves trees. If the ideas worked for you, we'd love to hear your success stories. Maybe we could turn them into 'Talking Heads'-style video or audio interviews on our website, so others can learn from you. That's one of the reasons why we wrote this book. So talk to us. *feedback@Smart-Questions.com*

Got a book you need to write?

Maybe you are a domain expert with knowledge locked up inside you. You'd love to share it and there are people out there desperate for your insights. But you don't think you are an author and don't know where to start. Making it easy for you to write a book is part of the Smart Questions Philosophy.

Let us know about your book idea, and let's see if we can help you get your name in print.

potentialauthor@Smart-Questions.com

Notes pages

We hope that this book has inspired you and that you have already scribbled your thoughts all over it. However if you have ideas that need a little more space then please use these notes pages.

Notes pages